NOETHERIAN RINGS
and their applications

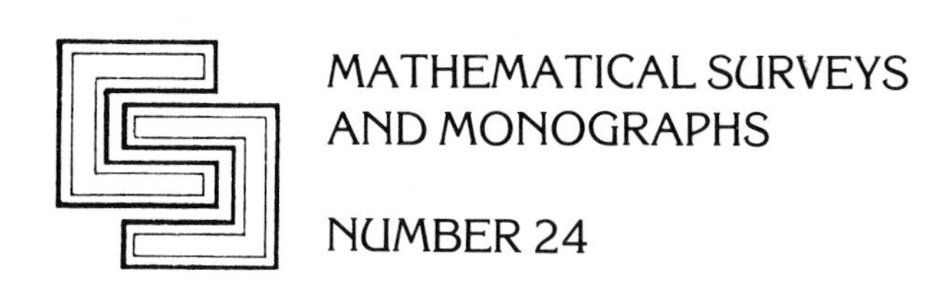

MATHEMATICAL SURVEYS AND MONOGRAPHS

NUMBER 24

NOETHERIAN RINGS and their applications

LANCE W. SMALL, EDITOR

American Mathematical Society
Providence, Rhode Island

1980 *Mathematics Subject Classification* (1985 *Revision*). Primary 16A33; Secondary 17B35, 16A27.

Library of Congress Cataloging-in-Publication Data

Noetherian rings and their applications.
(Mathematical surveys and monographs, 0076-5376; no. 24)
Largely lectures presented during a conference held at the Mathematisches Forschungsinstitut Oberwolfach, Jan. 23–29, 1983.
Includes bibliographies.
1. Noetherian rings–Congresses. 2. Universal enveloping algebras–Congresses. 3. Lie algebras–Congresses. I. Small, Lance W., 1941– . II. Mathematisches Forschungsinstitut Oberwolfach. III. Series.
QA251.4.N64 1987 512'.55 87-14997
ISBN 0-8218-1525-3 (alk. paper)

Contents

Preface

Noetherian rings abound in nature! Universal enveloping algebras of finite-dimensional Lie algebras and group algebras of polycyclic-by-finite groups are, for example, two wide classes of Noetherian rings. Yet, until Goldie's theorem was proved about thirty years ago, the "Noetherianness" of various types of noncommutative rings was not really effectively exploited.

Goldie's results provide the link between Noetherian rings and the much more studied case of Artinian rings. For instance, if R is a prime, right Noetherian ring, then R has a "ring of fractions" $Q(R)$ which is of the form D_n, $n \times n$ matrices over D a division ring. This last integer n is called the *Goldie rank* of R and plays an important role in some of the lectures in this collection. Later on the reader will meet some results involving Goldie ranks called "additivity principles" which arose, first, in studying rings satisfying a polynomial identity and then in enveloping algebras. For the moment, let's just state a very special case of such a result for semisimple Artinian rings: If $D_n \supset \Delta_{m_1}^{(1)} \oplus \cdots \oplus \Delta_{m_t}^{(t)}$, where D and the Δ's are division rings, then $n = \sum_{j=1}^{t} z_j m_j$ where the z's are *positive* integers. All sorts of useful and, occasionally, tantalizing generalizations will be seen in the lectures of Stafford and Jantzen.

Five of the six expository lectures collected in this volume were presented at a conference on Noetherian rings at the Mathematisches Forschungsinstitut, Oberwolfach during the week of January 23–29, 1983. The other article by Thomas J. Enright is an introduction to the representation theory of finite-dimensional semisimple Lie algebras which he gave at a London Mathematical Society Durham conference in July, 1983. Enright's paper will serve as an excellent complement to the surveys of Jantzen and Rentschler. One of the purposes of the Oberwolfach meeting was to bring together specialists in Noetherian ring theory and workers in other areas of algebra who use Noetherian methods and results. The meeting was organized by Professors Walter Borho and Alex Rosenberg and me.

If there is any common theme in these lectures, it is the study of the prime and primitive ideal spectra of various classes of rings: enveloping algebras, group algebras, polynomial identity rings, etc. A particularly useful tool is the Goldie rank of the ring factored by the relevant primes. Such investigations lead to the "additivity principles" discussed in Stafford's article. The Goldie rank then figures into the development of a convincing theory of localization in noncommutative rings which is described at the end of Stafford's paper.

The contributions of Jantzen and Rentschler treat the problem of classifying the primitive ideals of the universal enveloping algebras of finite-dimensional Lie algebras; Jantzen treats the semisimple case while Rentschler surveys the general situation. These papers are rather technical and the reader might also want to look at [1], for example.

Many of the methods developed for studying enveloping algebras can be successfully applied to the study of rings of differential operators on algebraic varieties and filtered algebras, more generally. J.-E. Björk provides a careful development of certain aspects of filtered Noetherian rings culminating in an easy-to-understand proof of Gabber's theorem on the integrability of the characteristic variety. Björk also provides background on rings of differential operators which have recently been a subject of great interest to ring theorists and analysts alike.

The concluding essay by D. R. Farkas focuses on the structure of group rings of polycyclic-by-finite groups (still the only known examples of Noetherian group algebras) discussing the Nullstellensatz, prime ideals à la Roseblade and those parts of the subject where Noetherian methods have proved valuable.

We turn now to some suggestions for additional reading on some of the more recent developments in Noetherian ring theory. A subject touched on in many of the lectures is affine algebras (i.e., algebras finitely generated over a field). The study of these algebras has grown immensely in recent years. A useful tool for investigating such algebras is Gelfand–Kirillov dimension. A clear, complete account of G–K dimension can be found in [4].

Localization theory continues to develop; there have been a number of "pay-offs," see, for example, [2] and [3]. Jategaonkar's monograph [5] is now available and provides an indispensable, though occasionally idiosyncratic, treatment of the state-of-art in localization techniques. Borho's lectures [1] also give interesting interpretations of certain localization techniques in the context of enveloping algebras.

Finally, what must surely become the standard reference for Noetherian rings, Robson and McConnell, is about to appear [6].

Lance W. Small

REFERENCES

1. W. Borho, *A survey on enveloping algebras of semisimple Lie algebras, I*, Can. Math. Soc. Conference Proceedings, Vol. **5**, AMS, Providence, 1986.

2. K. A. Brown and R. B. Warfield, Jr., *Krull and global dimensions of fully bounded Noetherian rings*, Proc. Amer. Math. Soc. **92** (1984), 169–174.

3. K. R. Goodearl and L. W. Small, *Krull versus global dimension in Noetherian PI rings*, Proc. Amer. Math. Soc. **92** (1984), 175–178.

4. G. R. Krause and T. H. Lenagan, *Growth of algebras and Gelfand–Kirillov dimension*, Research Notes in Mathematics **116**, Pitman, London, 1985.

5. A. V. Jategaonkar, *Localization in Noetherian rings*, London Math. Soc. Lecture Notes, Series **98**, Cambridge, 1986.

6. J. C. McConnell and J. C. Robson, *Rings with maximum condition*, John Wiley and Sons, to appear.

The Goldie Rank of a Module

J. T. STAFFORD

Let R be a semiprime, right Goldie ring, with semisimple Artinian right quotient ring $Q(R)$. If M is a right R-module, then the *Goldie rank* of M, written $\operatorname{Grk}_R M$, or $\operatorname{Grk} M$ if the context is clear, is just the length of the $Q(R)$-module $M \otimes_R Q(R)$. Equivalently, $\operatorname{Grk} M$ is the uniform dimension of M modulo its torsion submodule. The basic results about quotient rings and uniform dimension can be found, for example, in [8] or [11].

Stemming as it does from Goldie's Theorem, the Goldie rank of a module is one of the most classical concepts in the theory of Noetherian rings. However, there have been a number of recent results that have made effective use of the concept, and these form the basis of this article. In particular, one can frequently replace arguments involving localization—which even when available tend to be rather subtle—by arguments utilizing Goldie rank—which tend to be more straightforward and more widely applicable.

A nice illustration of this is given in Section 1, where we discuss variants of the Joseph–Small additivity principle. Given a ring R of finite index in a prime Noetherian ring S, this result relates $\operatorname{Grk}_S S$ to the $\operatorname{Grk} R/P$ as P ranges through the minimal prime ideals of R. The original proof required strong conditions on R, but by using Goldie rank, one obtains a very general result.

Similar applications of Goldie rank—in the guises of reduced rank and affiliated primes—are given in Section 2.

The Forster–Swan Theorem gives a bound for the number of generators, $g(M)$, of a module over a commutative Noetherian ring R in terms of the local number of generators $g_{R_P}(M_P)$ where P runs through the maximal ideals of R. In Section 3, we discuss noncommutative generalizations of this result. These have applications to classical algebraic K-theory, but also provide a rather useful continuity theorem that relates the Goldie rank of a ring to those of its prime factor rings.

In the fourth section, we turn to the question of when one can localize at a prime or semiprime ideal and discuss some of the recent work in this direction.

1. Additivity principles. The starting point of this section is the Joseph–Small Theorem. We denote the Gelfand–Kirillov dimension of a module M over a ring R by $\mathrm{d}(M)$. The basic properties of this dimension can be found in [3].

THEOREM 1.1 [19]. *Let $R \subset S$ be Noetherian rings such that S is a prime ring and that S is finitely generated as a left and a right R-module. Suppose that* $\mathrm{d}(R) < \infty$. *Then*

$$\mathrm{Grk}_S S = \sum r_i \mathrm{Grk}(R/P_i)$$

where $r_i \in \mathbf{Z}^+$ and the P_i run through the minimal prime ideals of R.

The crucial point is, of course, that the r_i are strictly positive integers. The *raison d'être* for the theorem lies in its application to the study of primitive ideals of enveloping algebras of semisimple Lie algebras, and some of these applications can be found, for example, in Jantzen's article [14] in this volume, or in [18].

Joseph–Small's original proof of Theorem 1.1 has three steps. First, if R is Artinian, then the result follows from an easy manipulation of idempotents and so the idea of the proof is to reduce the problem to the Artinian case. This requires two further steps. If I is an ideal of a ring R, define $\mathcal{C}(I)$ to be the elements of R that become regular in R/I. Then the second step is to show that R has an Artinian quotient ring; i.e., to show that $\mathcal{C} = \mathcal{C}_R(0)$ is an Ore set. The final step is to show that the module $S_{\mathcal{C}}$ actually equals the quotient ring $Q(S)$; equivalently, one must show that $\mathcal{C}$ is an Ore set of *regular* elements of S.

The final two steps are delicate questions, which make it difficult to generalize Joseph–Small's proof. In this section we give a rather different proof, due basically to Warfield [30] that shows that Theorem 1.1 is part of a very general result. Indeed, Warfield's result also works for non-Noetherian rings where it is easy to give counterexamples to the statement of Theorem 1.1.

EXAMPLE 1.2. Let

$$R = \begin{pmatrix} k & k[x] \\ 0 & k[x] \end{pmatrix} \subset S = M_2(k[x]).$$

In this case R does not have the Artinian quotient ring, although the conclusion of Theorem 1.1 does hold.

EXAMPLE 1.3 (Small). Let R_1 be the polynomial ring $k[x_1, \ldots]$ in infinitely many variables over a field and ϕ the k-endomorphism defined by $\phi(x_1) = 0$ and $\phi(x_i) = x_{i-1}$ for $i > 1$. Set

$$R = \left\{ \begin{pmatrix} a & x_1 b & 0 \\ c & d & 0 \\ 0 & 0 & \phi(d) \end{pmatrix} : a, b, c, d \in R_1 \right\} \subset S = M_3(R_1).$$

Then it is easy to see that R is a prime Goldie ring, with $\mathrm{Grk}\, R = 2$ and that S is a finitely generated R-module on either side. Thus, the conclusion of Theorem

1.1 fails for this ring. The reason is that

$$\begin{pmatrix} 1 & 0 & 0 \\ 0 & x_1 & 0 \\ 0 & 0 & 0 \end{pmatrix} \in \mathcal{C}_R(0) - \mathcal{C}_S(0).$$

Thus, if $\mathcal{C} = \mathcal{C}_R(0)$ then $S_\mathcal{C} \neq Q(S)$; indeed $S_\mathcal{C}$ is not even a ring.

We now turn to Warfield's result and begin with two easy lemmas.

LEMMA 1.4. *Let $R \subset S$ be simple Artinian rings. Then* $\operatorname{Grk} R$ *divides* $\operatorname{Grk} S$.

PROOF. This is an easy exercise—see [19, Lemma 3.8].

REMARK. A slight generalization of Lemma 1.4 will be useful later. Suppose that R is a prime, right Goldie subring of a simple Artinian ring S. If S is torsion-free as a right R-module, then regular elements of R remain left regular (and therefore regular) in S. Thus $Q(R)$ embeds in S and, by the lemma, $\operatorname{Grk} R$ divides $\operatorname{Grk} S$.

LEMMA 1.5. *Let R, S be rings with R prime Goldie. Suppose that M is an (S, R)-bimodule such that M is finitely generated as a left S-module.*

(i) *If M is torsion as a right R-module, then* $\text{r-ann}_R(M) = \{f \in R : Mf = 0\} \neq 0$.

(ii) *If M is torsion-free as a right R-module and S is simple Artinian, then M is naturally a right $Q(R)$-module.*

PROOF. (i) Write $M = \sum_1^n Sm_i$ for some $m_i \in M$. Since M is torsion, $\text{r-ann}_R(m_i)$ is an essential right ideal of R for each i. Since R is right Goldie, $\text{r-ann}_R(M) = \bigcap_1^n \text{r-ann}(m_i) \neq 0$. The proof of (ii) is left to the reader.

THEOREM 1.6. *Let $R \subset S$ be rings such that S is prime Goldie and R/P is Goldie for every prime ideal P of R. Then there exists a finite set of prime ideals $X = \{P_i\}$, including all the minimal prime ideals of R, such that*

$$\operatorname{Grk}_S S = \sum \{z_i \operatorname{Grk} R/P_i \ : \ P_i \in X \text{ and } z_i \in \mathbf{Z}^+\}.$$

PROOF. Since $Q(S)$ is an Artinian left $Q(S)$-module, we may choose a composition series

$$Q(S) = M_n \supset M_{n-1} \supset \cdots \supset M_0 = 0$$

of $(Q(S), R)$-bimodules. Write $P_i = \text{r-ann}_R(M_i/M_{i-1})$ and $X = \{P_i : 1 \leq i \leq n\}$. The simplicity of M_i/M_{i-1} as a bimodule ensures that each P_i is a prime ideal. Further, since $P_n P_{n-1} \cdots P_1 \subseteq Q(S) P_n P_{n-1} \cdots P_1 = 0$, each minimal prime ideal of R must be contained in X.

The key observation is that we may apply Lemma 1.5. For, let N be the torsion submodule of M_i/M_{i-1} as a right R/P_i-module. Then N is a $(Q(S), R/P_i)$-bimodule, and so, by Lemma 1.5(i), $N \neq M_i/M_{i-1}$. Since M_i/M_{i-1} is simple as a bimodule, N must therefore be zero. Thus, by Lemma 1.5(ii), M_i/M_{i-1} is a right $Q(R/P_i)$-module.

Thus, the action of $Q(R/P_i)$ on M_i/M_{i-1} provides an embedding of $Q(R/P_i)$ into $\mathrm{End}_{Q(S)} M_i/M_{i-1}$. So, by Lemma 1.4, there exists $z_i \in \mathbf{Z}^+$ such that

$$z_i \mathrm{Grk}\, Q(R/P_i) = \mathrm{Grk}(\mathrm{End}_{Q(S)}(M_i/M_{i-1})) = \mathrm{Grk}_{Q(S)}(M_i/M_{i-1}).$$

Finally, by the choice of our composition series,

$$\mathrm{Grk}\, Q(S) = \sum_1^n \mathrm{Grk}_{Q(S)}(M_i/M_{i-1}) = \sum_1^n z_i \mathrm{Grk}\, R/P_i.$$

The class of rings considered in the above result, rings R such that R/P is Goldie for every prime ideal P of R, is the natural place to consider many of the questions about Goldie rank. However, while it includes Noetherian rings and rings that satisfy a polynomial identity, we know of few other examples.

Theorem 1.6 shows that one can avoid the problems caused by S not being torsion-free as an R-module, simply by increasing the set of prime ideals of R that one is willing to consider. Thus, for example, if $R \subset S$ are the rings described by Example 1.3, set

$$P_1 = \begin{pmatrix} R_1 & xR_1 & 0 \\ R_1 & xR_1 & 0 \\ 0 & 0 & 0 \end{pmatrix} \subset R \quad \text{and} \quad M = \begin{pmatrix} 0 & 0 & R_1 \\ 0 & 0 & R_1 \\ 0 & 0 & R_1 \end{pmatrix} \subset S.$$

Then $P_1 = \text{r-ann}\, M_1$ and P_1 is a prime ideal of R. Furthermore, S/M_1 is torsion-free as a right R-module. Thus, in the notation of Theorem 1.6, $X = \{P_1, 0\}$ and $\mathrm{Grk}\, S = 1.\mathrm{Grk}\, R/P_1 + 1.\mathrm{Grk}\, R/0$.

In order to use Theorem 1.6 to derive Theorem 1.1, we need more information about the set X. Given an (S, R)-bimodule N for two rings S and R we say that N is a *torsion-free* (respectively, *finitely generated*) (S, R)-*bimodule*, if it is torsion-free (respectively, finitely generated) as a left S-module *and* as a right R-module.

PROPOSITION 1.7. *Keep the notation of Theorem* 1.6. *Let P_i be a minimal prime ideal of R and P_j any other element of X. Then there exists an R-bisubfactor N_{ij} of S such that N_{ij} is a torsion-free $(R/P_i, R/P_j)$-bimodule.*

Suppose, further, that RbR is Noetherian as a left and a right R-module, for all $b \in S$. Then N_{ij} may be chosen to be finitely generated on both sides.

PROOF. Fix j with $1 \le j \le n$ and let $C = M_j/M_{j-1}$ be as in the proof of the theorem. Then C is a finitely generated right module over the ring $V = \mathrm{End}_{Q(S)} C$. Thus, repeating the proof of the theorem, we obtain a composition series

$$C = C_m \supset \cdots \supset C_0 = 0$$

of (R, V)-bimodules such that, if $Y = \{Q_k : Q_k = \text{l-ann}_R(C_k/C_{k-1})\}$, then (i) Y contains every minimal prime ideal of R and (ii) C_k/C_{k-1} is torsion-free as a $(R/Q_k, V)$-bimodule. Notice that, as R/P_j embeds in V, each C_k/C_{k-1} remains torsion-free as a right R/P_j-module.

Now, by (i), $P_i = Q_k$ for some k. The first assertion of the Proposition will follow once we have shown that $C_k \cap S \neq C_{k-1} \cap S$, since $N_{ij} = C_k \cap S / C_{k-1} \cap S$ will then have the desired properties. Certainly, $M_j \cap S \neq M_{j-1} \cap S$, since the M_r are left ideals of $Q(S)$. Secondly, $U = \operatorname{End}_S(M_j \cap S / M_{j-1} \cap S)$ is a prime Goldie ring with quotient Artinian ring V (see [32, Theorem 3.3]). Set $D = (M_j \cap S + M_{j-1})/M_{j-1}$. Then, for each k, $C_k = (C_k \cap D)V$, and $C_k \cap D \supsetneqq C_{k-1} \cap D$, as required.

The final assertion of the proposition is a triviality—replace N_{ij} by RbR for any $b \neq 0 \in N_{ij}$.

Theorem 1.1 now follows from known properties of Gelfand–Kirillov dimension. Indeed, we have the following generalization.

COROLLARY 1.8 [2, Theorem 7.2]. *Let $R \subset S$ be rings such that S is prime Goldie and, for all $b \in S$, RbR is Noetherian as a left and as a right R-module. If* $\mathrm{d}(R) < \infty$, *then*

$$\operatorname{Grk} S = \sum \{ z_i \operatorname{Grk} R/P_i \ : \ P_i \text{ a minimal prime and } z_i \in \mathbf{Z}^+ \}.$$

Furthermore, $\mathrm{d}(R/P_i) = \mathrm{d}(R)$ *for all minimal primes P_i of R.*

PROOF. By [19, Lemma 2.5], $\mathrm{d}(R/Q) < \mathrm{d}(R)$ for any nonminimal prime ideal Q of R, while $\mathrm{d}(R) = \max\{\mathrm{d}(R/P) \ : \ P \text{ a minimal prime ideal of } R\}$. Thus, the corollary follows from Theorem 1.6 as soon as we have shown that $\mathrm{d}(R/P_i) = \mathrm{d}(R/P_j)$ for all P_i, $P_j \in X$. However, by Proposition 1.7 and [21, Corollary 3],

$$\mathrm{d}(R/P_i) = \mathrm{d}_{R/P_i} N_{ij} = \mathrm{d}_{R/P_j} N_{ij} = \mathrm{d}(R/P_j);$$

as required.

The formulation of Corollary 1.8 is particularly convenient for applications to enveloping algebras of Lie algebras. (We remark that, throughout this article, a Lie algebra will mean a finite-dimensional Lie algebra over a field k of characteristic zero.) For, let $\mathfrak{h}$ be a sub-Lie algebra of a Lie algebra $\mathfrak{g}$ with enveloping algebras $U(\mathfrak{h})$ and $U(\mathfrak{g})$, respectively. Then certainly $\mathrm{d}(U(\mathfrak{h})) < \infty$. Furthermore, an elementary argument, using the Poincaré–Birkhoff–Witt Theorem, shows that $U(\mathfrak{h})bU(\mathfrak{h})$ is a Noetherian $U(\mathfrak{h})$-module for any $b \in U(\mathfrak{g})$. Thus, Corollary 1.8 is applicable. Indeed, it is a natural generalization of Weyl's character formula. For, consider the special case when $\mathfrak{h}$ is a split Cartan subalgebra of a complex, semisimple Lie algebra $\mathfrak{g}$. Let P be the annihilator of a simple, finite-dimensional $U(\mathfrak{g})$-module V and take $R = U(\mathfrak{h})/P \cap U(\mathfrak{h}) \subset S = U(\mathfrak{g})/P$. In the notation of Corollary 1.8, $\operatorname{Grk} S = \dim_k V$, the P_i correspond to the weights of V and the z_i are the corresponding multiplicities. Thus, the conclusion of Corollary 1.8 is indeed Weyl's character formula. This observation comes from [2, Section 9] where other applications of the corollary are mentioned.

It is not clear to what extent the hypotheses of Corollary 1.8 can be weakened without destroying the conclusion. It seems likely that some version of the condition that RbR be Noetherian is going to be necessary. However, the condition

$d(R) < \infty$ could be removed entirely if one could answer the following question in the affirmative.

PROBLEM 1.9. *Let R be a prime Noetherian ring and M a finitely generated R-bimodule. If M is torsion-free as a left R-module, show that M cannot be torsion as a right R-module.*

Although this is known to be true for many of the standard classes of rings, it is a long-standing question as to whether it is true for all Noetherian rings. Variations on this problem will trouble us throughout this survey and we will discuss it more fully in Section 5.

We end this section by noting that Lemma 1.4 remains true if Goldie rank is replaced by PI degree. The proof of Theorem 1.6 can then be used to prove the following result.

PROPOSITION 1.10 [1, Theorem 7.3(a)]. *Let R be a subring of a prime PI ring S. Then there exists a finite set X of prime ideals of R, including all minimal prime ideals of R, such that*

$$\operatorname{PIdeg} S = \sum\{z_i \operatorname{PIdeg} R/P_i \ : \ P_i \in X \text{ and } z_i \in \mathbf{Z}^+\}.$$

It would be interesting to know if there exists an easy proof of the "Big Bergman–Small Theorem" [1, Theorem 6.8], which relates the PI degree of a prime PI ring to that of its factor rings.

2. Affiliated primes and reduced rank. The trick used in Section 1 to avoid a "difficult" localization by reducing to the case of torsion-free modules over prime rings has been used in various ways, and we mention two of these in this section. We assume throughout that R is a ring in which every prime factor ring is Goldie (although we will usually add on some sort of Noetherian condition).

First, there are affiliated series. Given a right R-module M then an *affiliated series* for M (if it exists) is a chain of submodules

$$M = M_n \supset \cdots \supset M_0 = 0$$

such that, for each i, $P_i = \text{r-ann}\, M_i/M_{i-1}$ is maximal among right annihilators of submodules of M/M_{i-1} and $M_i = \{m \in M \ : \ mP_i \subseteq M_{i-1}\}$. The set $\{P_i \ : \ 1 \leq i \leq n\}$ is called the *affiliated set of primes* of M corresponding to this series.

Obviously, such a series exists if M is a finitely generated module over a right Noetherian ring R. However, there are other circumstances when they arise. For example, suppose that M is also a finitely generated, torsion-free left module over a left Goldie ring S. If I is an ideal of R and $N = \{m \in M \ : \ mI = 0\}$ then it is routine to check that N is a left S-submodule of M such that M/N is torsion-free over S. Thus, we may use the uniform dimension of ${}_SM$ to construct the affiliated series for M_R. As an application, one can prove Theorem 1.6 by taking an affiliated series for S as a right R-module, rather than a composition series for $Q(S)$.

One useful application of affiliated series is to determine the regular elements of a Noetherian ring. This is illustrated by the next lemma. Given an ideal I of a ring R, write $'\mathcal{C}(I)$ for the set of elements of R that become left regular in R/I.

LEMMA 2.1. *Let R be a Noetherian ring and $R = M_n \supset \cdots \supset M_0 = 0$ an affiliated series for R_R, with affiliated set of primes $\{P_i = \text{r-ann}\, M_i/M_{i-1}\}$. Then $\bigcap \mathcal{C}(P_i) \subseteq' \mathcal{C}(0)$.*

PROOF. We need R to be left Noetherian, since this means we can apply Lemma 1.5(i) to show that each M_i/M_{i-1} is a torsion-free right R/P_i-module.

Suppose that $c \in \bigcap \mathcal{C}(P_i)$ and $a \in R$ are such that $ac = 0$. Now $a \in M_i$ for some i. However M_i/M_{i-1} is a torsion-free right R/P_i-module and $c \in \mathcal{C}(P_i)$. Since $ac \in M_{i-1}$, this forces $a \in M_{i-1}$. Induction now completes the proof.

Unfortunately, the stronger statement $'\mathcal{C}(0) = \bigcap \mathcal{C}(P_i)$ need not hold in Lemma 2.1 (see [24, Section 3]). However, by "patching together" several affiliated series, one can prove:

THEOREM 2.2 [23, Corollary 2.3]. *Let R be a Noetherian ring. Then there exists a finite set of prime ideals $\{Q_i \ : \ 1 \leq i \leq n\}$ of R such that $\mathcal{C}(0) = \bigcap \mathcal{C}(Q_i)$.*

A number of related results may be found in [23] and [24]. We note that Theorem 2.2 and most of the other results in [23] and [24] will fail for right Noetherian rings. Essentially, this is because one cannot apply Lemma 1.3(i).

To complement Theorem 2.2, we have the following result, which can be thought of as a sort of Chinese Remainder Theorem. The proof is left to the reader.

PROPOSITION 2.3 [24, Proposition 2.4]. *Let $\{Q_1, \ldots, Q_n\}$ be a set of prime ideals of R and K a right ideal of R such that, for each i, $K \cap \mathcal{C}(Q_i) \neq \emptyset$. Then there exists $k \in K$ such that $k \in \bigcap \mathcal{C}(Q_i)$.*

Let $N(R)$ stand for the nilradical of a ring R, and $J(R)$ for the Jacobson radical. If R has an Artinian quotient ring Q, then Theorem 2.2 follows from Small's Theorem, since now $\mathcal{C}(0) = \mathcal{C}(N(R))$. In this case, another obvious invariant for an R-module M is the length of the Q-module $M \otimes_R Q$. This again is a concept that generalizes to an arbitrary Noetherian ring.

Let M be a finitely generated right module over a Noetherian ring R. Then

$$\rho_R(M) = \sum_0^\infty \mathrm{Grk}_{R/N(R)} MN(R)^i/MN(R)^{i+1}$$

is a finite integer and is called the *reduced rank* of M. The following facts about reduced rank are easy to ascertain and generalize the corresponding facts about the length of a module over an Artinian ring.

(2.4) If $0 \to M_1 \to M \to \overline{M} \to 0$ is a short exact sequence of right R-modules, then $\rho(M) = \rho(M_1) + \rho(\overline{M})$. In particular, given a chain of submodules $M = M_k \supset \cdots \supset M_0 = 0$ such that each M_i/M_{i-1} is killed by a minimal prime ideal P_i, then $\rho(M) = \sum \operatorname{Grk}_{R/P_i} M_i/M_{i-1}$.

(2.5) $\rho(M) = 0$ if and only if, for each $m \in M$ there exists $c \in \mathcal{C}(N(R))$ such that $mc = 0$.

We do not wish to spend much time on reduced rank, since we have nothing to add to the results in [8] or [9]. However, reduced rank does seem to provide very easy proofs of a number of the results from the literature. To illustrate this, let us give a proof of the "hard" part of Small's Theorem.

THEOREM 2.6. *Let R be a Noetherian ring. Then R has an Artinian quotient ring if and only if $\mathcal{C}(0) = \mathcal{C}(N(R))$.*

PROOF. [9] Suppose that $\mathcal{C}(0) = \mathcal{C}(N(R))$. Let $c \in \mathcal{C}(0)$ and $a \in R$. Since c is regular, $R \simeq cR$ and $\rho(R) = \rho(cR)$. Thus, by 2.4, $\rho(R/cR) = 0$ and hence, $\rho(aR + cR/cR) = 0$. By 2.5, this means that $ac' \in cR$ for some $c' \in \mathcal{C}(N(R)) = \mathcal{C}(0)$. Hence $\mathcal{C}(0)$ is indeed a right (and left) Ore set. The fact that $Q = R_{\mathcal{C}(0)}$ is Artinian follows from the observation that, given right ideals $I \subsetneqq J$ of Q, then $\rho_R(J \cap R/I \cap R) \neq 0$.

Another application of reduced rank is to give an easy proof of Jategaonkar's Principal Ideal Theorem [15]. Indeed, one has the following result.

THEOREM 2.7 [9, Theorem 4.6]. *Let X be an invertible ideal of a prime Noetherian ring R; that is, there exists an R-sub-bimodule X^{-1} of $Q(R)$ such that $XX^{-1} = X^{-1}X = R$. Then any prime ideal minimal over X has height one.*

The proof of this result amounts to calculating the reduced rank of various R/X-modules.

3. Generating modules efficiently and a continuity theorem. The aim of this section is to try to determine—or at least bound—the minimal number of generators, $g(M)$, of a finitely generated right module M over a Noetherian ring R. This turns out to depend largely upon the Goldie ranks of certain factor modules of M. In the process, we are forced to prove a rather surprising result about the Goldie ranks of prime factor rings of R.

Our first aim is to find an appropriate generalization of the Forster–Swan Theorem.

THEOREM 3.1 [10, 29]. *Let R be a commutative Noetherian ring and M a finitely generated R-module. Then*

$$\begin{aligned} g(M) &\leq \max_{P \text{ prime}} \{g_{R_P}(M_p) + K\dim R/P\} \\ &\leq \max_{P \text{ maximal}} \{g_{R_P}(M_P)\} + K\dim R. \end{aligned}$$

One way to view this result is that it gives an upper bound for $g(M)$ that is "trivial" to determine. For, by Nakayama's Lemma, $g_{R_P}(M_P)$ is just the dimension of the vector space M_P/M_PP_P over the field R_P/P_P.

Before we can generalize this result, we need to remove reference to the local rings R_P, so that at least the statement is meaningful over a noncommutative ring. This again follows from Nakayama's Lemma, since

$$\begin{aligned} g_{R_P}(M_P) &= g_{R_P/P_P}(M_P/M_PP_P) \\ &= g_{Q(R/P)}(M/MP \otimes Q(R/P)). \end{aligned}$$

This final term makes sense in general. Thus, for a prime ideal P in an arbitrary Noetherian ring R, and a finitely generated right R-module M, write

$$g(M,P) = g_{Q(R/P)}(M/MP \otimes Q(R/P)).$$

As in the commutative case, this is trivial to determine. Indeed, since $Q(R/P)$ is simple Artinian, $M/MP \otimes Q(R/P) \simeq Q(R/P)^{(m)} \oplus I$ for some integer m and cyclic $Q(R/P)$-module I. Thus, we define

$$\hat{g}(M,P) = (\operatorname{Grk} M/MP)/(\operatorname{Grk} R/P).$$

Then, $g(M,P)$ is the smallest integer greater than or equal to $\hat{g}(M,P)$.

Given a right module M over a Noetherian ring R, let $K\dim_R M$ stand for the Krull dimension of M in the sense of Rentschler and Gabriel [12]. In particular, $K\dim R$ stands for the Krull dimension of R as a right R-module. With this notation, we can state the noncommutative generalization of the first part of the Forster–Swan Theorem.

THEOREM 3.2 [25, Theorem 3.1]. *Let M be a finitely generated right module over a Noetherian ring R. Then*

$$g(M) \leq \max_{P\,\text{prime}} \{g(M,P) + K\dim R/P\}.$$

The proof of this result is difficult. However, it becomes considerably easier when the ring is fully bounded Noetherian (FBN), and we will outline the proof for this case. This case of the theorem is due to Warfield [31]. The reader is reminded that a prime ring is *bounded* if every essential right ideal contains a two-sided ideal and that a ring is *fully bounded* if every prime factor ring is bounded. The obvious example of a fully bounded ring is a PI ring (and, in fact, there are not too many other examples).

If M is a finitely generated right module over an FBN ring R, then we will actually prove

$$g(M) \leq \max_{\substack{P\,\text{prime} \\ g(M,P)\neq 0 \\ J(R/P)=0}} \{g(M/P) + K\dim R/P\}. \tag{3.3}$$

This is achieved by induction on the right-hand side. Since R is FBN, the RHS is zero if and only if $M = 0$, and this starts the induction. The general step has two parts.

(i) The RHS of 3.3 is attained at only finitely many prime ideals.

(ii) Suppose that $P_1, \ldots, P_n$ are prime ideals of R such that $g(M, P_i) > 0$ for each i. Then there exists $m \in M$ such that $g(M/mR, P_i) < g(M, P_i)$ for $1 \leq i \leq n$.

The second statement is a generalization of the Chinese Remainder Theorem (Proposition 2.3), and the proof is routine. Part (i) is an easy consequence of the following result.

PROPOSITION 3.4 [31, Theorem 4]. *Let M be a finitely generated right module over an* FBN *ring R, and P a prime ideal of R. Then there exists an ideal $I \supsetneqq P$ such that $\hat{g}(M, P) = \hat{g}(M, Q)$ whenever $Q \supseteq P$ but $Q \not\supseteq I$.*

REMARK. This result can be rephrased to say that $\hat{g}(M, ?)$ is locally constant on $\operatorname{Spec} R$ endowed with the *patch topology*. In this topology, a basic open neighbourhood of $P \in \operatorname{Spec} R$ is a set of the form $\{Q \supseteq P \ : \ Q \not\supseteq I$ for some fixed ideal I with $I \supsetneqq P\}$.

Outline of Proof. We may assume that $P = 0$ and (with a little work) that M is torsion-free. Replacing M by $M^{(n)}$ for some integer n, we may further assume that $\operatorname{Grk} R$ divides $\operatorname{Grk} M$. Thus, M embeds in a free module, say $V \simeq R^{(r)}$, such that V/M is torsion. Similarly, there exists a free submodule $W \simeq R^{(r)}$ of M such that M/W is also torsion. Take $I = \text{r-ann}\, V/W$, which, since R is FBN, is a nonzero ideal. If $Q \not\supseteq I$, then $V/VQ + M$, being a homomorphic image of $V/V(Q + I)$, is a torsion R/Q-module. Similarly, $M/MQ + W$ is torsion. Thus

$$\begin{aligned} r = \hat{g}(V, Q) &= \hat{g}(M + VQ/VQ, Q) \leq \hat{g}(M, Q) \\ &= \hat{g}(MQ + W/MQ, Q) \leq \hat{g}(W, Q) = r. \end{aligned}$$

Hence, $\hat{g}(M, Q) = r = \hat{g}(M, 0)$; as required.

If a prime ideal P satisfies $J(R/P) = 0$ then it is an intersection of primitive ideals. Since a primitive ideal is maximal in an FBN ring, Proposition 3.4 and (3.3) also give the second inequality of the Forster–Swan Theorem.

COROLLARY 3.5. *If M is a finitely generated right module over an* FBN *ring R, then*

$$g(M) \leq \max_{P \text{ maximal}} \{g(M, P)\} + K \dim R.$$

Let us now return to an arbitrary Noetherian ring and Theorem 2.2, and note two complications that arise in the proof. First, the statement of (3.3) will no longer hold; if M is a torsion module over a simple Noetherian ring, then the right-hand side of (3.3) is zero. One gets around this by bringing the term $K \dim M$ into one's induction. Secondly, Proposition 3.4 will also not hold (we will give an example later) and so one needs to use the following weaker result.

PROPOSITION 3.6. *Let M be a finitely generated module over a Noetherian ring R. Pick a prime ideal P of R and some $\varepsilon > 0$. Then there exists an ideal $I = I(\varepsilon, P) \supsetneqq P$ such that*

$$|\hat{g}(M, P) - \hat{g}(M, Q)| < \varepsilon$$

whenever $Q \supseteq P$ *but* $Q \not\supseteq I$.

Equivalently, $\hat{g}(M,?)$ *is continuous in the patch topology.*

Similarly, Corollary 3.5 will not hold in general; consider the case when R is a prime ring with exactly one nonzero ideal, say P, and M is the direct sum of a large number of copies of P. However, only a slight modification is needed to make it valid.

COROLLARY 3.7. *Let* M *be a finitely generated module over a Noetherian ring* R. *Then*

$$g(M) \leq \max_{P \text{ primitive}} \{g(M,P)\} + K \dim R.$$

The results of [25] and [31] are stronger than those mentioned above, as they give information about the *stable number of generators*, $s(M)$, of the module M. This is defined as follows: $s(M)$ is the least integer s such that, given any $r \geq s$ and $m_1, \ldots, m_{r+1} \in M$ such that $M = \sum_1^{r+1} m_i R$, then there exist $\lambda_1, \ldots, \lambda_r \in R$ such that $M = \sum_1^r (m_i + m_{r+1})\lambda_i R$. All the results of this section can be improved by replacing $g(M)$ by $s(M)$. This has applications to the classical algebraic K-theory of Noetherian rings. For example, a special case of (the improved) Theorem 3.2 is that $s(R) \leq 1 + K \dim R$, which is a noncommutative version of Bass's Stable Range Theorem. Another application is:

THEOREM 3.8 [25, Section 5]. *Let* R *be a Noetherian ring and* M *a projective, finitely generated right* R*-module. Suppose that* $\hat{g}(M,P) \geq 1 + K \dim R$ *for every prime ideal* P *of* R. *Then* $M \simeq M' \oplus R$ *for some module* M'. *Further, if* $M \oplus R \simeq N \oplus R$ *for some module* N, *then* $M \simeq N$.

Further applications can be found in [25] and [31]. One problem with the results is worth mentioning. In the commutative case, the Forster–Swan Theorem (which is actually slightly stronger than Theorem 3.1) is the best possible result. It is not known if this remains true in the noncommutative case. For example, if M is a finitely generated, torsion module over a simple Noetherian ring R, then Theorem 3.2 says that $g(M) \leq 1 + K \dim R$, and this can be improved to $g(M) \leq 1 + K \dim M$. Unfortunately, there is no known example where $g(M) > 1$. It is known that one can always embed M into a cyclic module [27, Corollary 3.5]. One problem is that there seem to be no techniques for finding lower bounds for $g(M)$ in the noncommutative case, other than the triviality $g(M) \geq \{\max g(M,P) \ : \ P \text{prime}\}$.

For the rest of this section, we will discuss Propositions 3.4 and 3.6, as we feel that they should be rather useful results, even though the present applications are rather limited. Unfortunately, the conclusion of Proposition 3.4 does not hold for an arbitrary Noetherian ring. For example, let R be the first Weyl algebra over the integers; so $R = A_1(\mathbf{Z}) = \mathbf{Z}[x,y \ : \ xy - yx = 1]$. For any prime q of $\mathbf{Z}$, the ideal $P_q = qR + x^q R$ is a prime ideal and $\bigcap P_q = 0$. Now x is certainly regular in R, but not regular modulo any P_q. In other words, $\hat{g}(R/xR, 0) = 0$

but $\hat{g}(R/xR, P_q) > 0$ for all q. Thus, the conclusion of Proposition 3.4 will not hold. Indeed, this seems to happen remarkably often, as counterexamples are also provided by the enveloping algebra $U(\mathfrak{sl}_2)$ and the group ring $\mathbf{Z}G$ of the group $G = \langle a, b \ : \ aba^{-1}b^{-1} = z, \ z \text{ central}\rangle$.

One application of Proposition 3.6 is the following.

COROLLARY 3.9. *Let R be a prime Noetherian ring and $n > 0$ an integer. Then there exists an ideal $I = I(n) \neq 0$ such that, if P is a prime ideal of R with $P \not\supseteq I$, then either*

(i) $\operatorname{Grk} R$ *divides* $\operatorname{Grk} R/P$, *or*

(ii) $\operatorname{Grk} R/P \geq n$.

If R is FBN, *then* (ii) *can be deleted.*

REMARK. In the special case $n = 1$ the Corollary shows that

$$\bigcap\{P \in \operatorname{Spec} R \ : \ \operatorname{Grk} R/P < \operatorname{Grk} R\} \neq 0.$$

First Proof. Let M be a uniform right ideal of R and apply Proposition 3.6 with $\varepsilon = 1/(n \operatorname{Grk} R)$. Then there exists an ideal $I \neq 0$ such that

$$|\hat{g}(M, Q) - \hat{g}(M, 0)| < \varepsilon$$

for any prime ideal Q with $Q \not\supseteq I$. Now use the fact that $\hat{g}(M, 0) = 1/\operatorname{Grk} R$ and that $\hat{g}(M, Q)$ is an integer multiple of $1/\operatorname{Grk}(R/Q)$ to deduce the corollary.

Second Proof. Since there exists an easy direct proof of this result, we will give it here. The proof was discovered jointly with R. Snider. In the proof we only need that R is a prime right Noetherian ring.

Suppose that the corollary is false. Then, for some integer n not divisible by $\operatorname{Grk} R$, there exist prime ideals $\{P_i \ : \ i \in I\}$ such that $\bigcap P_i = 0$ but $\operatorname{Grk} R/P_i = n$ for each i. For $r \neq 0 \in R$, let $I_r = \{i \in I \ : \ r \notin P_i\}$, and let $\mathcal{F}$ be an ultrafilter containing the $\{I_r \ : \ r \neq 0 \in R\}$. Since $\bigcap P_i = 0$, clearly R embeds in the ultraproduct $S = (\prod R/P_i)/\mathcal{F}$. Further, for each i, $Q(R/P_i) \simeq M_n(D_i)$ for some division ring D_i. Since the ultraproduct $D = (\prod D_i)/\mathcal{F}$ is also a division ring [13, Proposition 2.1, p. 77], it follows that S is a prime right Goldie ring with quotient ring $T = M_n(D)$. In particular, $\operatorname{Grk} S = n$. It only remains to show that T is a torsion-free, right R-module, since then the additivity principle will imply that $\operatorname{Grk} R$ divides $n = \operatorname{Grk} T$ (see the remark after Lemma 1.4).

Thus, suppose that the torsion submodule, say V, of T_R is nonzero. Let $J = \text{r-ann}_R V$. By Lemma 1.5(i), J is a nonzero ideal. Write $J = \sum_1^n a_i R$, for some $a_i \in R$. Pick $v \neq 0 \in V \cap S$ and let $\prod[v_j + P_j]$ be a preimage of v in $\prod R/P_j$. For each i, $va_i = 0$ and so $\Omega_i = \{j \ : \ v_j a_i \in P_j\} \in \mathcal{F}$. Thus, $\Omega = \bigcap \Omega_i \in \mathcal{F}$. Observe that $\Omega = \{j \ : \ v_j J \subseteq P_j\}$. However, since $v \neq 0$, $\Psi_1 = \{j \ : \ v_j \notin P_j\} \in \mathcal{F}$ and, as $J \neq 0$, $\Psi_2 = \{j \ : \ J \not\subseteq P_j\} \in \mathcal{F}$. Thus $\Psi_1 \cap \Psi_2 \in \mathcal{F}$. Since J is an ideal, $\Psi_1 \cap \Psi_2 = I \backslash \Omega$; which is absurd.

It would be interesting to know whether there exists a similarly easy proof of Proposition 3.6. The above proof can be so generalized if one knows that $\Omega = \{\operatorname{Grk} R/P \ : \ P \in \operatorname{Spec} R\}$ is bounded above. Unfortunately, when Ω

is unbounded, then the corresponding ultraproduct $S = (\prod R/P_i)/\mathcal{F}$ is rather unpleasant.

As was true of Proposition 3.6, Corollary 3.7 will fail in an arbitrary Noetherian ring if one deletes possibility (ii) of the corollary. The example involves taking idealizers inside $U(\mathfrak{sl}_2)$ and can be found in [26].

An application of Proposition 3.6, similar to Corollary 3.9, occurs when one considers linked prime ideals. Two prime ideals P and Q in a Noetherian ring R are said to have a *second layer link*, written $P \rightsquigarrow Q$, if there exists an ideal A, with $PQ \subseteq A \subsetneqq P \cap Q$, such that $P \cap Q/A$ is a torsion-free $(R/P, R/Q)$-bimodule. The terminology will become clear in the next section when we consider the problems of localization, but for the moment we just regard links as a subtlety in the ideal structure of the ring. It is not known whether comparable primes $P \subseteq Q$ can be linked, and this is closely related to Problem 1.9.

An easy example of links is provided by the two-dimensional, complex, solvable Lie algebra $\mathfrak{g}$; so $\mathfrak{g} = \mathbf{C}x + \mathbf{C}y$ with $[x, y] = x$. For, let $P_n = xU(\mathfrak{g}) + (y + n)U(\mathfrak{g})$, for $n \in \mathbf{Z}$. Then it is readily checked that $P_n \rightsquigarrow P_{n+1}$.

One consequence of Proposition 3.6 is that there cannot be too many linked prime ideals.

COROLLARY 3.10. *Let P be a prime ideal of the Noetherian ring R. Then*

(i) *Suppose that R is* FBN *(or that Proposition* 3.4 *holds for R). Then P is linked $P \rightsquigarrow Q$ to only finitely many prime ideals Q.*

(ii) *In general, for any integer n, P is linked to only finitely many prime ideals Q with* $\operatorname{Grk} R/Q \leq n$.

We begin with a reduction, the proof of which is identical to that of [17, Theorem 6.2.11].

LEMMA 3.11. *For any subset Y of* $\operatorname{Spec} R$, *the following are equivalent:*

(a) *P is linked to only finitely many prime ideals Q in Y.*

(b) *For any prime ideal $T \subsetneqq P$, let $Y_T = \{Q \in Y : Q \supseteq T$ and $P/T \rightsquigarrow Q/T\}$. Then $\bigcap\{Q : Q \in Y_T\} \supsetneqq T$.*

Proof of Corollary 3.10. The lemma immediately implies that there exist at most finitely many prime ideals $Q \supsetneqq P$ such that $P \rightsquigarrow Q$. For, part (b) of the lemma is automatically satisfied when $Y = \{Q : Q \supsetneqq P$ and $P \rightsquigarrow Q\}$.

Set $Y = \{Q \in \operatorname{Spec} R : Q \not\supseteq P\}$ in case (i) of the corollary, and $Y = \{Q \in \operatorname{Spec} R : Q \not\supseteq P$ and $\operatorname{Grk} R/Q \leq n\}$ in case (ii). If the corollary is false, then by the lemma, there exists a prime ideal $T \subsetneqq P$ such that $\bigcap\{Q : Q \in Y_T\} = T$. Thus, replacing R by R/T and Y by Y_T, we may assume that $\bigcap\{Q : Q \in Y\} = 0$ and that $P \rightsquigarrow Q$ for all $Q \in Y$.

We can now rephrase the definition of a link in terms of Goldie ranks. Suppose that $Q \in Y$. Since $Q \not\supseteq P$,

$$\hat{g}(P, Q) = \hat{g}(P/P \cap Q, Q) + \hat{g}(P \cap Q/PQ, Q) \geq 1 + 1/\operatorname{Grk}(R/Q). \tag{3.12}$$

In case (ii) of the Corollary, apply Proposition 3.6 with $\varepsilon = 1/(2n)$. Then there exists an ideal $I \neq 0$ such that, if $Q \not\supseteq I$, then

$$\hat{g}(P, Q) \leq \hat{g}(P, 0) + \varepsilon = 1 + \varepsilon.$$

Since $\bigcap\{Q : Q \in Y\} = 0$, this contradicts (3.12). Case (i) of the Corollary is proved by using Proposition 3.4 in place of 3.6.

Given a prime ideal P in a Noetherian ring R, define the *clique* of P to be the equivalence class of P in $\operatorname{Spec} R$, defined by $\rightsquigarrow$. By Corollary 3.11 and its left-hand analogue, we have:

COROLLARY 3.13. *The clique of P is a countable set.*

The obvious question raised by the last two results is whether cliques are locally finite; that is, must a prime ideal P be linked, $P \rightsquigarrow Q$, to at most finitely many prime ideals Q? Although this is the case for many of the standard classes of rings (for example, enveloping algebras of solvable Lie algebras, group rings of polycyclic by finite groups [5] and, of course, FBN rings) in general, the answer is no (see [26]).

4. Localization. Having spent much of this article on ways of avoiding localization, in this section we will discuss when one *can* localize at a prime or semiprime ideal. Most of this section is distilled from Jategaonkar's memoir [17]. An ideal I of a ring R is *(right) localizable* if $\mathcal{C}(I)$ is a (right) Ore set. I is *classically right localizable* if, moreover, the Jacobson radical $J(R_I)$ of the localized ring R_I satisfies the Artin–Rees property. We remark that it is unknown whether these two concepts are equivalent in a Noetherian ring. However, the Artin–Rees property is a useful extra property. (For example, it implies that the Jacobson conjecture holds for R_I.)

Let us note two obvious cases when one cannot (right) localize at a prime ideal P of a Noetherian ring: first, if there exists a second layer link $Q \rightsquigarrow P$ for some prime ideal $Q \neq P$ (as, for example, happens in the enveloping algebra of the 2-dimensional, solvable Lie algebra); second, when P is the unique nonzero ideal of a prime ring R (as, for example, happens in appropriate factor rings of enveloping algebras of nonsolvable Lie algebras). Let us give an *ad hoc* proof of the first case, where to make life easy we will assume that $Q \not\subseteq P$. So there exists an ideal $A \subsetneq Q \cap P$ with $Q \cap P/A$ torsion-free as an $(R/Q, R/P)$-bimodule. Pick $q \in Q \cap \mathcal{C}(P)$ and $a \in (Q \cap P) \backslash A$. If P is right localizable, then there exists $q' \in \mathcal{C}(P)$ and $a' \in R$ such that $qa' = aq'$. As $q \in \mathcal{C}(P)$ but $a \in P$, certainly $a' \in P$. Thus, $qa' \in QP \subseteq A$. Thus, $aq' \in A$, which contradicts the fact that $Q \cap P/A$ is torsion-free as a right R/P-module. This proof can also be modified to work in the second example. (Use the fact that there exists an essential extension of R/P by a faithful, torsion R-module M with $MP = M$.) Alternatively, of course, one could just note that Nakayama's Lemma would fail in the localization R_P.

In both cases, the obstruction is really caused by a bad essential extension of R/P. So, let $E(R/P)$ be the injective hull of R/P as a right R-module, and write $\mathrm{ann}_{E(R/P)}P = \{f \in E(R/P) \ : \ fP = 0\}$ and $X_P = E(R/P)/\mathrm{ann}_{E(R/P)}(P)$. Then, in the first example, R/Q appears as a submodule of X_P, and in the second case, the faithful torsion module appears as a submodule of X_P. (This also explains the terminology "second layer link" : the link $Q \rightsquigarrow P$ occurs in X_P—the second layer of $E(R/P)$). Surprisingly, these two obstructions are the only obstacles to localization and (at least when cliques are finite) one can even bypass the problem caused by linked primes, by localizing at the intersection of the members of the clique.

THEOREM 4.1 [17, Theorem 7.3.1]. *Let S be a semiprime ideal in a Noetherian ring R. Then the following are equivalent:*

(i) *S is classically right localizable.*

(ii) (a) *If P is a prime ideal minimal over S and $Q \rightsquigarrow P$ for some prime ideal Q, then Q is minimal over S.*

(b) *Suppose that P is a prime ideal minimal over S and that A is a finitely generated submodule of X_P with a prime annihilator T. Then $\mathrm{Grk}_{R/T}(A) > 0$.*

REMARKS. (i) There is a similar result for (nonclassical) right localization.

(ii) A subset Ω of $\mathrm{Spec}\, R$ is said to satisfy the *right second layer condition* if condition (ii)(b) holds for every prime ideal P in Ω. As usual, R satisfies the right second layer condition if this condition holds for $\Omega = \mathrm{Spec}\, R$.

(iii) A somewhat weaker form of link is the following. Two prime ideals are *ideal linked* if there exist ideals $A \subsetneqq B$ such that B/A is a torsion-free $(R/P, R/Q)$-bimodule. The theorem also holds with this definition of link, simply because the two concepts coincide when the second layer condition is satisfied (see [17, Theorem 8.2.4]). However, the two concepts are *not* the same in general, as can be shown by looking at appropriate subrings of $U(\mathfrak{sl}_2)$ [26].

(iv) A finite clique Ω is called a *clan* if $S = \bigcap\{Q \ : \ Q \in \Omega\}$ is localizable. Of course, by the theorem, S is then the unique, largest, localizable, semiprime ideal contained in Q, for any $Q \in \Omega$.

(v) It is easy to construct a prime ideal P in a Noetherian ring R such that P is right, but not left, localizable. For example, consider the ring R of upper 2×2 matrices over a field. However, it is not known if this is still possible in a prime ring.

It is easy to find rings for which the second layer condition fails. Generalizing one of the examples from the beginning of the section, we have:

LEMMA 4.2. *Let R be a prime Noetherian ring with a unique, minimal, nonzero ideal, say I. Then the second layer condition fails (on both sides) for R.*

PROOF. Use the fact that $E_R(R/I)$ is a torsion R-module with $\mathrm{ann}_R E(R/I) = 0$.

In particular, the second layer condition will fail for the enveloping algebra of any nonsolvable Lie algebra (in characteristic zero) and for most rings constructed as idealizers. However, these are really the only standard examples of Noetherian rings where it does fail.

THEOREM 4.3. *If R is one of the following rings, then R satisfies the second layer condition:*

(i) R *is* FBN.

(ii) $R = kG$ *is the group ring of a polycyclic-by-finite group over a commutative, Noetherian ring* k [4, Theorem 4.2] *or* [17, Theorem A.4.6].

(iii) $R = U(\mathfrak{g})$ *is the enveloping algebra of a solvable Lie algebra* [4, Theorem 4.4] *or* [17, Theorem A.3.9].

Apart from its relevence to localization, the second layer condition has a number of other useful consequences; essentially because it forces the ring to have a generous supply of two-sided ideals. Some of these applications are described in the next section.

Let R be a Noetherian ring that satisfies the second layer condition. Then, by Theorem 4.1, a prime ideal P is minimal over a localizable, semiprime ideal S if and only if the clique of P is finite. Unfortunately, cliques are often infinite. For example, if R is the enveloping algebra of a solvable Lie algebra $\mathfrak{g}$ over an algebraically closed field, then a clique will either consist of a single prime ideal, or will be infinite [6, Theorem 2.13]. Furthermore, there will exist at least one infinite clique when $\mathfrak{g}$ is nonnilpotent. This final comment follows from the fact that it is true for the two-dimensional, solvable Lie algebra; see the comments before Corollary 3.10.

However, there is some evidence that one may still be able to localize at an infinite clique Ω in Spec R, and obtain a ring that satisfies many of the properties of a local ring. Thus, one wants to invert $\mathcal{C}(\Omega) = \bigcap\{\mathcal{C}(Q) : Q \in \Omega\}$ and this immediately raises a problem. Given $a \in R$ and $c \in \mathcal{C}(\Omega)$, set $K = \{f \in R : af \in cR\}$. As in the case of a finite clique, one can show that, for each $Q \in \Omega$, $K \cap \mathcal{C}(Q) \neq \emptyset$. However, one now needs an "infinite Chinese Remainder Theorem" to deduce that $K \cap \mathcal{C}(\Omega) \neq \emptyset$.

The next lemma covers most of the cases where such a result is known.

LEMMA 4.4. *Let R be a Noetherian ring such that R contains an uncountable, central field k. Let Ω be a countable set of prime ideals of R and suppose that either* (i) $\{\operatorname{Grk} R/Q : Q \in \Omega\}$ *is bounded, or* (ii) R *is* FBN.

Let K be a right ideal of R such that, for all $Q \in \Omega$, there exists $c \in K \cap \mathcal{C}(Q)$. Then there exists $d \in K$ with $d \in \mathcal{C}(\Omega)$.

REMARK. (i) The lemma fails if one does not assume some sort of uncountable subring. For example, the result fails if Ω consists of all but one of the maximal ideals of $\mathbf{Q}[x, y]$ and K is the remaining maximal ideal. However, it is not known if the lemma is true when Ω is a clique in an arbitrary Noetherian ring R.

(ii) If $K = \sum_1^r a_i R$, then some element of the form $d = a_1 + \sum_2^r a_i \lambda_i$, for $\lambda_i \in R$, will satisfy the conclusion of the lemma.

PROOF. We may assume that Ω is infinite and, by replacing R by $R/\bigcap\{Q : Q \in \Omega\}$, that $\bigcap\{Q : Q \in \Omega\}$ is zero. By induction, we may assume that the result is true in any factor ring of R. It is easy to check that K must now be an essential right ideal of R and so contain a regular element, say c. Once again, Propositions 3.4 and 3.6 imply the existence of an ideal $I \neq 0$ such that $c \in \mathcal{C}(P)$ for every prime ideal $P \not\supseteq I$. By induction, there exists $b \in K$ such that $b \in \bigcap\{\mathcal{C}(Q) : Q \supseteq I \text{ and } Q \in \Omega\}$. Thus, for any $Q \in \Omega$, either $a \in \mathcal{C}(Q)$ or $b \in \mathcal{C}(Q)$. It follows that $a + bu \in \mathcal{C}(Q)$ for all but finitely many $u \in k$ (use the proof that eigenvectors are linearly independent!). Finally, as k is uncountable, there must exist at least one $u \in k$ such that $a + bu \in \mathcal{C}(\Omega)$.

PROPOSITION 4.5. *Suppose that Ω is a clique in a Noetherian ring R and that Ω satisfies the second layer condition. Suppose further that R contains an uncountable central field and that either R is* FBN *or that $\{\mathrm{Grk}\, R/Q : Q \in \Omega\}$ is bounded. Then $\mathcal{C}(\Omega)$ is an Ore set.*

PROOF. By Corollary 3.13, Ω is countable, and so Lemma 4.4 can be applied. The rest of the proof is similar to that of Theorem 4.1 and forms [17, Theorem 7.2.15].

For affine PI rings, this result is due to Müller [22], and for enveloping algebras of solvable Lie algebras (where all prime ideals are completely prime) it is due to Brown [7]. By another, unpublished, result of Brown, $\{\mathrm{Grk}\, R/Q : Q \in \Omega\}$ is bounded when R is the group ring of a polycyclic by finite group, and so Proposition 4.5 can also be applied in this case. It is not known whether the result holds when $\{\mathrm{Grk}\, R/Q : Q \in \Omega\}$ is unbounded.

The consequences of localizing at an infinite clique have not been fully explored, although the resulting localization $S = R_{\mathcal{C}(\Omega)}$ is known to have some useful properties. For example:

(i) Every primitive ideal of S has the form QS for some $Q \in \Omega$.

(ii) S/QS is simple Artinian for each $Q \in \Omega$.

(iii) A finitely generated, essential extension of a simple S-module is Artinian.

(iv) S has stable range one (this follows easily from Remark (ii) to Lemma 4.4).

5. Bimodules. It is becoming increasingly apparent that two-sided Noetherian rings have much more pleasant properties than one-sided Noetherian rings, even if one is only concerned with one-sided properties. However, the question of how one should effectively use the two-sided Noetherian condition is still a major problem. One of the subsidiary themes in this article has been the use of torsion-free, finitely generated bimodules, and these clearly provide a useful method for understanding how the property of being left Noetherian affects the right module structure of a Noetherian ring. However, here again problems abound. Obviously Problem 1.9 is one of the more important questions, but more vaguely one can raise:

QUESTION. Let R, S be prime Noetherian rings and M a finitely generated, torsion-free (R, S)-bimodule. How are the properties of R and S (or of ${}_RM$ and M_S) related?

For example, one can ask if $K\dim R = K\dim S$ or if $K\dim({}_RM) = K\dim(M_S)$. The only general result in this direction is the result of Lenagan [20] that shows that this is true for Krull dimension zero; i.e., ${}_RM$ is Artinian if and only if M_S is Artinian. As can be seen from say [8, Chapters 4, 5, 13], this has proved to be a very useful result. A related question is: given an Artinian right R-module A, is $A \otimes_R M$ an Artinian S-module? Even if $M = S \supset R$, this seems intractable.

The situation is much more pleasant if one assumes the second layer condition. Let us try to indicate why this is so. Suppose that R is a prime Noetherian ring satisfying the second layer condition, and that P is a nonzero prime ideal of R. If M is a finitely generated, essential extension of R/P, then M must be a torsion R-module. So, by the second layer condition, $\operatorname{ann}_R M \neq 0$. This provides a large number of ideals of R, since $0 = \operatorname{ann}_R E(R/P) = \bigcap\{\operatorname{ann} M : M$ a finitely generated, essential extension of $R/P\}$. Now, suppose that N is a torsion-free, finitely generated (S, R)-bimodule for some ring S. Then the same considerations show that there exist bisubmodules $N_1 \subset N_2$ of N such that $P = \text{r-ann}\, N_2/N_1$ (see [16, Lemma 1.2]). By looking at the annihilators in S of such bisubfactors, one sees that S also has "many" ideals. In particular, if the classical Krull dimension, cl-Kdim, of a ring is defined in terms of chains of prime ideals, one obtains that $\text{cl-Kdim}\, S \geq \text{cl-Kdim}\, R$. The first part of the next result is an immediate consequence of this observation.

PROPOSITION 5.1. [16]. *Let R be a Noetherian ring that satisfies the second layer condition.*

(i) *Let P and Q be prime ideals of R such that there exists a finitely generated, torsion-free $(R/P, R/Q)$-bimodule M. Then Q cannot strictly contain P.*

(ii) *The prime ideals in a clique are incomparable.*

(iii) *Jacobson's conjecture holds; that is $\bigcap J(R)^n = 0$.*

(iv) *R has primary decomposition.*

We remark that (iv) fails for an appropriate factor of $U(\mathfrak{sl}_2)$ [4, Example 6.4]. The status of the other parts of the proposition is open for an arbitrary Noetherian ring. We remark that classical Krull dimension will not in general "pass through" a bimodule. For example, one can construct, for any integer n, a simple Noetherian domain R and an idealizer subring S of R such that S is Noetherian with $\text{cl-Kdim}\, S = n$. In this case, R is itself a finitely generated, torsion-free (S, R)-bimodule. Note that a simple ring automatically satisfies the second layer condition.

Thus, in trying to understand the sort of problems mentioned in this section, one might do well to look at the most obvious class of rings that does not satisfy the second layer condition: enveloping algebras of semisimple Lie algebras, and the rings that can be derived from them. The reader will have observed that $U(\mathfrak{sl}_2)$ has been used (intentionally) to produce a number of the counterexamples mentioned in this article.

Added in Proof. Since this survey was written, there have been several developments on the questions raised here. An easy proof of Proposition 3.6 now exists—see [33]. When the survey was written, [17] only existed as a preprint. The published version now includes most of the results on infinite cliques that are proved here. Proposition 4.5 has been proved independently by Warfield in [36]. In that paper, Warfield also gives a more detailed description of the rings obtained by localizing at infinite cliques. In [34] it is shown that infinite cliques are always localizable in affine Noetherian PI rings. Thus in that case, one need not assume that the ring contains an uncountable field. One of the questions raised in Section 5 has also been answered; there exists an Artinian module M over a Noetherian domain R and a finite overring S of R such that $M \otimes S$ is not Artinian. See [35] for this and related examples.

REFERENCES

1. G. M. Bergman and L. W. Small, *P.I. degrees and prime ideals*, J. Algebra **33** (1975), 435–462.
2. W. Borho, *On the Joseph-Small additivity principle for Goldie ranks*, Comp. Math. **57** (1982), 3–29.
3. W. Borho and H. Kraft, *Über die Gelfand-Kirillov-Dimension*, Math. Ann. **220** (1976), 1–24.
4. K. A. Brown, *Module extensions over Noetherian rings*, J. Algebra **69** (1981), 247–260.
5. K. A. Brown, *The structure of modules over polycyclic groups*, Math. Proc. Camb. Phil. Soc. **89** (1981), 257–283.
6. K. A. Brown, *Localisation, bimodules and injective modules for enveloping algebras of solvable Lie algebras*, Bull. Sci. Math. **107** (1983), 225–251.
7. K. A. Brown, *Ore sets in enveloping algebras*, Comp. Math. **53** (1984), 347–367.
8. A. W. Chatters and C. R. Hajarnavis, *Rings with chain conditions*, Research Notes in Math. No. 44, Pitman, Boston, 1980.
9. A. W. Chatters, A. W. Goldie, C. R. Hajarnavis, and T. H. Lenagan, *Reduced rank in Noetherian rings*, J. Algebra **61** (1979), 582–589.
10. O. Forster, *Über die Anzahl der Erzengenden eines Ideals in einen Noethershen Ring*, Math. Z. **84** (1964), 80–87.
11. A. W. Goldie, *The structure of Noetherian rings*, in Lecture Notes in Math., Vol. 246, Springer-Verlag, Berlin, 1972.
12. R. Gordon and J. C. Robson, *Krull dimension*, Mem. Amer. Math. Soc., Vol. **133** (1973).
13. N. Jacobson, *Basic Algebra II*, W. H. Freeman, San Francisco, 1980.
14. J. C. Jantzen, *Primitive ideals in the enveloping algebra of a semisimple Lie algebra*, this volume.
15. A. V. Jategaonkar, *Relative Krull dimension and prime ideals in right Noetherian rings*, Comm. in Algebra **4** (1974), 429–468.
16. A. V. Jategaonkar, *Noetherian bimodules, primary decomposition and Jacobson's conjecture*, J. Algebra **71** (1981), 379–400.
17. A. V. Jategaonkar, *Localisation in Noetherian rings*, Lecture Note Series No. 98, London Math. Soc., Cambridge Univ. Press, Cambridge, 1986.

18. A. Joseph, *Goldie rank in the enveloping algebra of a semisimple Lie algebra*, I, II, III, J. Algebra **65** (1980), 269–283; 284–306; **73** (1981), 295–326.

19. A. Joseph and L. W. Small, *An additivity principle for Goldie rank*, Israel J. Math. **31** (1978), 105–114.

20. T. H. Lenagan, *Artinian ideals in Noetherian rings*, Proc. Amer. Math. Soc. **51** (1975), 499–500.

21. T. H. Lenagan, *Gelfand-Kirillov dimension and affine PI rings*, Comm. in Algebra **10** (1982), 87–92.

22. B. J. Müller, *Two-sided localisation in Noetherian PI rings*, in Ring Theory **44** (Lecture Notes in Pure and Applied Math. vol. 51), Marcel Dekker, New York, 1979.

23. L. W. Small and J. T. Stafford, *Regularity of zero-divisors*, Proc. London Math. Soc. **44** (1982), 405–419.

24. J. T. Stafford, *Noetherian full quotient rings*, Proc. London Math. Soc. **44** (1982), 385–404.

25. J. T. Stafford, *Generating modules efficiently: algebraic K-theory for noncommutative Noetherian rings*, J. Algebra **69** (1981), 312–346; Corrig. ibid. **82** (1983), 294–296.

26. J. T. Stafford, *On the ideals of a Noetherian ring*, Trans. Amer. Math. Soc. **289** (1985), 381–392.

27. J. T. Stafford, *Modules over prime Krull rings*, J. Algebra **95** (1985), 332–342.

28. J. T. Stafford, *Generating modules efficiently over noncommutative rings* in Séminaire Dubreil-Malliavin, (Lecture Notes in Math. No. 924), Springer-Verlag, 1982.

29. R. W. Swan, *The number of generators of a module*, Math. Z. **102** (1967), 318–322.

30. R. B. Warfield, *Prime ideals in ring extensions*, J. London Math. Soc. **28** (1983), 453–460.

31. R. B. Warfield, *The number of generators of a module over a fully bounded ring*, J. Algebra **66** (1980), 425–447.

32. J. M. Zelmanowitz, *Endomorphism rings of torsion-less modules*, J. Algebra **5** (1967), 325–341.

33. K. R. Goodearl, *Patch continuity of normalised ranks of modules over one-sided Noetherian rings*, Pacific J. Math. **122** (1986), 83–94.

34. B. J. Müller, *Affine Noetherian PI rings have enough clans*, J. Algebra **97** (1985), 116–129.

35. J. T. Stafford, *Non-holonomic modules over Weyl algebras and enveloping algebras*, Inventiones Math. **79** (1985), 619–638.

36. R. B. Warfield, *Noncommutative localised rings*, to appear.

Representation theory of semisimple Lie algebras

THOMAS J. ENRIGHT

1. Introduction. This article is an account of three lectures given at Durham. These lectures covered some of the background material which would be needed in the later lectures during the conference. Most, if not all, of this material has now become standard in representation theory.

In section two we summarize some of the structural properties of semisimple Lie algebras over an algebraically closed field. We include the root space decomposition, the Chevalley restriction theorem and the Harish–Chandra isomorphism for the center of the enveloping algebra. Section three includes a discussion of Verma modules and the translation principle. The last section describes the Zuckerman derived functors. The functors are the principal link between the category of highest weight modules and the category of Harish–Chandra modules. This material is not as well known as that included in sections two and three and so here we have sketched many of the proofs.

I wish to thank Professors Goldie and McConnell for the invitation to visit Durham and participate in their conference.

2. Basic structural facts. Here we summarize the basic facts for semisimple Lie algebras over an algebraically closed field F of characteristic zero. There are several excellent texts where this material is presented. For clarification or further developments the reader should consult [3], [6], and [9].

Let $\mathfrak{g}$ be a finite-dimensional Lie algebra. For any finite-dimensional vector space V, let $\mathrm{End}(V)$ denote the linear endomorphisms of V. $\mathrm{End}(V)$ is a Lie algebra with commutator as Lie bracket. A *representation* π of $\mathfrak{g}$ on V is a Lie algebra homomorphism of $\mathfrak{g}$ into $\mathrm{End}(V)$. When convenient we shall suppress the homomorphism π and call V a $\mathfrak{g}$-module. For $x \in \mathfrak{g}$, define $\mathrm{ad}(x)(y) = [x, y]$, $y \in \mathfrak{g}$. The Jacobi identity for the Lie bracket on $\mathfrak{g}$ implies that ad is a representation of $\mathfrak{g}$ on $\mathfrak{g}$. This representation is called the *adjoint representation.*

Let V be a $\mathfrak{g}$-module and β a bilinear form on V. We say β is invariant if $\beta(x \cdot e, f) + \beta(e, x \cdot f) = 0$ for all $x \in \mathfrak{g}$ and $e,\ f \in V$. The Killing form $\langle x, y\rangle = \mathrm{trace}(\mathrm{ad}(x)\mathrm{ad}(y))$ is an invariant form on $\mathfrak{g}$ which is an essential tool in the decomposition of semisimple Lie algebras.

THEOREM 2.1. *The following are equivalent:*

(i) *The radical of* $\mathfrak{g}$ *is* 0.
(ii) *The Killing form is nondegenerate.*
(iii) $\mathfrak{g}$ *is a product of simple Lie algebras.*

A Lie algebra having one of these properties is called semisimple.

THEOREM 2.2. (Weyl) *Assume* $\mathfrak{g}$ *is semisimple and* π *is a finite-dimensional representation of* $\mathfrak{g}$. *Then* π *is completely reducible.*

Humphrey's text [**6**] presents a nice proof of this classical result based on two facts. The first is the existence of the Casimir elements (associated with nondegenerate invariant forms on $\mathfrak{g}$) and the second is the fact that $\mathfrak{g} = [\mathfrak{g}, \mathfrak{g}]$, which implies that all one-dimensional representations are trivial.

In the representation theory of compact Lie groups, the maximal tori are the essential subgroups. The analogue for Lie algebras is the Cartan subalgebra (CSA). Suppose $\mathfrak{g}$ is semisimple; then a CSA $\mathfrak{h}$ of $\mathfrak{g}$ is a maximal abelian subalgebra with all elements semisimple (i.e., for $x \in \mathfrak{h}$, $\mathrm{ad}(x)$ is diagonalizable).

THEOREM 2.3. *Assume* $\mathfrak{g}$ *is semisimple. Then* $\mathfrak{g}$ *admits Cartan subalgebras and any two CSA's are conjugate by an automorphism of* $\mathfrak{g}$.

For the remainder of this section assume $\mathfrak{g}$ is semisimple. We now describe the *root space decomposition* of $\mathfrak{g}$. Fix $\mathfrak{h}$ a CSA of $\mathfrak{g}$. By definition $\mathrm{ad}(x)$ is diagonalizable for all $x \in \mathfrak{h}$. Moreover, since $\mathfrak{h}$ is abelian these operators commute. So we can simultaneously diagonalize the adjoint action of $\mathfrak{h}$ on $\mathfrak{g}$. Let $\mathfrak{h}^*$ be the algebraic dual of $\mathfrak{h}$. For $\alpha \in \mathfrak{h}^*$, let $\mathfrak{g}_\alpha = \{y \in \mathfrak{g} \mid [x, y] = \alpha(x)y, \forall x \in \mathfrak{h}\}$. For $\alpha \neq 0$, the nonzero $\mathfrak{g}_\alpha$ are called the root spaces and the α are called the roots. Let Δ denote the set of roots. We have:

$$\mathfrak{g} = \mathfrak{g}_0 \oplus \sum_{\alpha \in \Delta} \mathfrak{g}_\alpha.$$

THEOREM 2.4. $\mathfrak{h} = \mathfrak{g}_0$, $\dim \mathfrak{g}_\alpha = 1$ *and* $\alpha \in \Delta$ *implies* $-\alpha \in \Delta$.

The classification of semisimple Lie algebras proceeds along this path. We will not take this up, except to note that two semisimple Lie algebras are isomorphic if and only if their corresponding sets of roots are isomorphic as abstract root systems (cf. [**6**]).

Now write $\Delta = \Delta^+ \cup -\Delta^+$ and put $\mathfrak{n} = \sum_{\alpha \in \Delta^+} \mathfrak{g}_\alpha, \mathfrak{n}^- = \sum_{\alpha \in \Delta^+} \mathfrak{g}_{-\alpha}$. We call Δ^+ a system of positive roots for Δ if $\mathfrak{n}$ is a Lie subalgebra of $\mathfrak{g}$. Such systems always exist and we now fix one which we denote by Δ^+. This gives the *triangular decomposition* of $\mathfrak{g}$:

$$\mathfrak{g} = \mathfrak{n}^- \oplus \mathfrak{h} \oplus \mathfrak{n}. \tag{2.5}$$

Put $\mathfrak{b} = \mathfrak{h} \oplus \mathfrak{n}$. Then $\mathfrak{b}$ is called a *Borel subalgebra* of $\mathfrak{g}$. Any subalgebra containing a Borel subalgebra is called a *parabolic subalgebra* of $\mathfrak{g}$. These subalgebras play a fundamental role in the representation theory of $\mathfrak{g}$.

The Weyl group $\mathfrak{w} = \mathfrak{w}(\mathfrak{g}, \mathfrak{h})$ is the subgroup of $\mathrm{End}(\mathfrak{h}^*)$ generated by the reflections s_α, $\alpha \in \Delta$, where

$$s_\alpha(\lambda) = \lambda - 2\langle \alpha, \lambda\rangle / \langle \alpha, \alpha\rangle \cdot \alpha, \quad \lambda \in \mathfrak{h}^*. \tag{2.6}$$

For any finite-dimensional $\mathfrak{g}$-module V, $\mathfrak{h}$ acts semisimply. So V is the direct sum of the subspaces $V_\lambda = \{v \in V \mid x \cdot v = \lambda(x)v, \forall x \in \mathfrak{h}\}$. The $\lambda \in \mathfrak{h}^*$ with $V_\lambda \neq 0$ are called the weights of V, and V_λ is the weight space for λ. For any finite-dimensional representation of $\mathfrak{g}$, the set of weights will be stable under the action of the Weyl group.

For any Lie algebra $\mathfrak{a}$, let $U(\mathfrak{a})$ denote the universal enveloping algebra and $Z(\mathfrak{a})$ the center of $U(\mathfrak{a})$. For simple $\mathfrak{g}$-modules, $Z(\mathfrak{g})$ will act by scalars; and so, the structure of $Z(\mathfrak{g})$ will be a useful aid in representation theory. For $\mathfrak{g}$ semisimple, $Z(\mathfrak{g})$ is a polynomial algebra in $\dim \mathfrak{h}$-variables. In the remainder of this section we give a precise description of this subalgebra of $U(\mathfrak{g})$.

Let $S(\mathfrak{g})$ (resp. $S(\mathfrak{h})$) denote the algebra of polynomials on $\mathfrak{g}^*$ (resp. $\mathfrak{h}^*$) and let res be the restriction map from $\mathfrak{g}^*$ to $\mathfrak{h}^*$

$$\mathrm{res}\colon S(\mathfrak{g}) \to S(\mathfrak{h}). \tag{2.7}$$

Let G be the group of inner automorphisms of $\mathfrak{g}$. Denote by $I(\mathfrak{g})$ (resp. $I(\mathfrak{h})$) the G-invariant polynomials on $\mathfrak{g}^*$ (resp. $\mathfrak{w}$-invariant polynomials on $\mathfrak{h}^*$).

THEOREM 2.8. (Chevalley) *The map res induces an isomorphism*

$$I(\mathfrak{g}) \xrightarrow{\sim} I(\mathfrak{h}).$$

Since $\mathfrak{w}$ is isomorphic to the normalizer in G of $\mathfrak{h}$/ centralizer in G of $\mathfrak{h}$, res maps $I(\mathfrak{g})$ into $I(\mathfrak{h})$. Moreover, since the G-orbit of $\mathfrak{h}^*$ in $\mathfrak{g}^*$ is Zariski dense, this map is an injection. The surjectivity follows from detailed knowledge of the finite-dimensional representations of $\mathfrak{g}$.

The group G also acts on $U(\mathfrak{g})$ and the subalgebra of G-invariant elements is $Z(\mathfrak{g})$. We now give the analogue to (2.8) for $Z(\mathfrak{g})$. Let φ denote the projection of $U(\mathfrak{g})$ onto $U(\mathfrak{h})$ given by the decomposition $U(\mathfrak{g}) = U(\mathfrak{h}) \oplus (\mathfrak{n}^- U(\mathfrak{g}) + U(\mathfrak{g})\mathfrak{n})$. Since $\mathfrak{h}$ is abelian we identify $U(\mathfrak{h})$ and $S(\mathfrak{h})$ and let $I(\mathfrak{h})$ be as above. Let γ be the automorphism of $S(\mathfrak{h})$ which transforms the polynomial p on $\mathfrak{h}^*$ into the polynomial $\lambda \mapsto p(\lambda - \rho), \lambda \in \mathfrak{h}^*$.

THEOREM 2.9. (Harish-Chandra) *The map $\gamma \circ \varphi$ induces an isomorphism*

$$Z(\mathfrak{g}) \xrightarrow{\sim} I(\mathfrak{h}).$$

For any $\lambda \in \mathfrak{h}^*$, let χ_λ be the homomorphism of $Z(\mathfrak{g})$ into the field F given by $\chi_\lambda(z) = (\gamma \circ \varphi(z))(\lambda)$. The χ_λ will be called the characters of $Z(\mathfrak{g})$ or the infinitesimal characters of simple $\mathfrak{g}$-modules. The theorem implies that $\chi_\lambda = \chi_{\lambda'}$ if and only if λ and λ' lie in the same $\mathfrak{w}$-orbit.

3. Verma modules and finite-dimensional $\mathfrak{g}$-modules. Verma modules, their irreducible quotients and their annihilators have become basic objects in the representation theory of semisimple Lie algebras. Here we define them and summarize their basic properties. Following this we describe the action of translation functors on these modules. The reader should consult [**3**] and [**7**] for more on the theory of Verma modules and [**7**] and [**11**] for translation functors.

Let notation be as in section two. So, $\mathfrak{g} = \mathfrak{n}^- \oplus \mathfrak{b}$ and $\mathfrak{b} = \mathfrak{h} \oplus \mathfrak{n}$. For any $\lambda \in \mathfrak{h}^*$, let $\mathbb{C}_\lambda$ denote the one-dimensional $\mathfrak{b}$-module where $\mathfrak{h}$ acts by λ and $\mathfrak{n}$ acts by zero. Put $\rho = \frac{1}{2}\sum_{\alpha\in\Delta^+}\alpha$. The Verma module $M(\lambda)$ is defined to be the induced module:

$$M(\lambda) = U(\mathfrak{g}) \otimes_{U(\mathfrak{b})} \mathbb{C}_{\lambda-\rho}. \tag{3.1}$$

Since $\mathfrak{g} = \mathfrak{u}^- \oplus \mathfrak{b}$, as an $\mathfrak{h}$-module $M(\lambda)$ is isomorphic to $U(\mathfrak{u}^-) \otimes \mathbb{C}_{\lambda-\rho}$. Let 1_λ denote the canonical cyclic vector $1 \otimes 1$ in $M(\lambda)$. Recall from section two the homomorphism χ_λ. Then, for $z \in Z(\mathfrak{g})$, $z \cdot 1_\lambda = \chi_\lambda(z) \cdot 1_\lambda$. Since 1_λ is cyclic, $Z(\mathfrak{g})$ acts by χ_λ on all of $M(\lambda)$.

The maximal proper submodule of $M(\lambda)$ is the sum of all submodules which do not contain a weight vector of weight $\lambda - \rho$. Let $L(\lambda)$ denote the irreducible quotient of $M(\lambda)$ by this maximal submodule. $L(\lambda)$ is the unique irreducible quotient of $M(\lambda)$. The first fundamental relation for Verma modules was established by Bernstein, Gel′fand, and Gel′fand.

THEOREM 3.2 [1].
(a) *For* $\lambda, \mu \in \mathfrak{h}^*$, $\dim \operatorname{Hom}_{\mathfrak{g}}(M(\mu), M(\lambda))$ *is either zero or one.*
(b) *Assume* $\alpha \in \Delta^+$ *and* $2\langle\lambda,\alpha\rangle/\langle\alpha,\alpha\rangle \in \mathbb{N}$. *Then* $\dim \operatorname{Hom}_{\mathfrak{g}}(M(s_\alpha\lambda), M(\lambda)) = 1$ *and any nonzero map is an injection.*
(c) *All maps between Verma modules are compositions of those described in* (b).

It is frequently convenient to view Verma modules as objects in the category $\mathcal{O}$. This category is the category of all finitely generated $U(\mathfrak{g})$-modules which are semisimple as $\mathfrak{h}$-modules and are $U(\mathfrak{b})$-locally finite. For $\lambda \in \mathfrak{h}^*$, let $\mathcal{O}_\lambda$ be the subcategory of $\mathcal{O}$ where all operators $z - \chi_\lambda(z)\cdot 1$ are locally nilpotent, $z \in Z(\mathfrak{g})$. Objects in these categories have formal characters defined as follows. For $A \in \mathcal{O}$ and $\mu \in \mathfrak{h}^*$, let A_μ denote the μ weight space in A. Define $\operatorname{ch} A$ to be the element of the group ring of $\mathfrak{h}^*$:

$$\operatorname{ch} A = \sum_{\mu\in\mathfrak{h}^*} (\dim A_\mu) e^\mu. \tag{3.3}$$

Put

$$q = e^\rho \prod_{\alpha\in\Delta^+} (1 - e^{-\alpha}) = \prod_{\alpha\in\Delta^+} (e^{\alpha/2} - e^{-\alpha/2}). \tag{3.4}$$

Since $M(\lambda) \simeq U(\mathfrak{n}^-) \otimes \mathbb{C}_{\lambda-\rho}$ and $(1 - e^a)^{-1} = 1 + e^a + e^{2a} + \cdots$ we have:

$$q \cdot \operatorname{ch} M(\lambda) = e^\lambda. \tag{3.5}$$

Both $\operatorname{ch} M(w\lambda)$, $w \in \mathfrak{w}$ and $\operatorname{ch} L(w\lambda)$, $w \in \mathfrak{w}$ form bases for the Grothendieck group of $\mathcal{O}_\lambda$. The proof of the Kazhdan–Lusztig conjectures has given the precise relations between these two bases [**8**].

An element $\lambda \in \mathfrak{h}^*$ is called *integral* if $2\langle\lambda,\alpha\rangle/\langle\alpha,\alpha\rangle \in \mathbb{N}$ for all $\alpha \in \Delta^+$ and *dominant integral* if these integers are all nonnegative. The classical theorem of the highest weight can now be formulated.

THEOREM 3.6.

(a) (Cartan) *The map $\lambda \mapsto L(\lambda+\rho)$ induces a bijection from the set of dominant integral elements of $\mathfrak{h}^*$ onto the set of equivalence classes of irreducible finite-dimensional $\mathfrak{g}$-modules.*

(b) (Weyl) *For λ dominant integral,*

$$q \cdot \operatorname{ch} L(\lambda+\rho) = \sum_{w \in \mathfrak{w}} \det(w) e^{w(\lambda+\rho)}.$$

In the remainder of this section we describe some basic functors in semisimple representation theory, the translation functors. Here we describe them in the context of the category $\mathcal{O}$ where the theory is frequently regarded as the Jantzen translation principle. In the context of Harish–Chandra modules the functors are called the Zuckerman translation functors [**11**].

For any object $A \in \mathcal{O}$ there is a finite set μ_i, $1 \leq i \leq r$, with $A = \sum A_i$ and $A_i \in \mathcal{O}_{\mu_i}$. Let p_{μ_i} be the projection which maps A to A_i. So $p_\lambda \colon \mathcal{O} \to \mathcal{O}_\lambda$. Fix $\mu, \xi \in \mathfrak{h}^*$ with $\operatorname{Re}\langle\mu,\alpha\rangle \geq 0$ for all $\alpha \in \Delta^+$ and ξ dominant integral. Let F equal the finite-dimensional $\mathfrak{g}$-module $L(\xi+\rho)$. Now define exact functors

$$(3.7) \qquad \begin{cases} \varphi_\mu^{\mu+\xi}(A) = p_{\mu+\xi}(F \otimes p_\mu(A)) \\ \psi_{\mu+\xi}^{\mu}(A) = p_\mu(F^* \otimes p_{\mu+\xi}(A)). \end{cases}$$

When convenient we drop the sub and superscripts and write φ and ψ in (3.7). For fixed μ and ξ, φ and ψ are adjoint functors; i.e., for $A, B \in \mathcal{O}$, $\operatorname{Hom}_{\mathfrak{g}}(\varphi A, B) \xrightarrow{\sim} \operatorname{Hom}_{\mathfrak{g}}(A, \psi B)$ and $\operatorname{Hom}_{\mathfrak{g}}(\psi A, B) \xrightarrow{\sim} \operatorname{Hom}_{\mathfrak{g}}(A, \varphi B)$. The usefulness of these functors rests in their computability in $\mathcal{O}$.

PROPOSITION 3.8. *Let μ and ξ be as above. Also assume that μ is regular; i.e., $\langle\mu,\alpha\rangle \neq 0$ for all $\alpha \in \Delta$.*

(a) *The restriction of φ to $\mathcal{O}_\mu$ gives an equivalence of $\mathcal{O}_\mu$ and $\mathcal{O}_{\mu+\xi}$ with inverse ψ.*

(b) *For all $w \in \mathfrak{w}$,*

$$\varphi M(w\mu) = M(w(\mu+\xi)), \quad \psi M(w(\mu+\xi)) = M(w\mu),$$
$$\varphi L(w\mu) = L(w(\mu+\xi)), \quad \psi L(w(\mu+\xi)) = L(w\mu).$$

Since (a) follows from (b) and (b) has an elementary proof we sketch it here. Let $\nu = w\mu$ and $\nu' = w(\mu+\xi)$. Choose a flag of $\mathfrak{b}$-modules for $F \otimes \mathbb{C}_{\nu-\rho} = B_r \supset \cdots \supset B_0 = 0$ with $r = \dim F$ and $B_i/B_{i-1} \simeq \mathbb{C}_{\mu_i-\rho}$. Note that $F \otimes M(\nu) \simeq U(\mathfrak{g}) \otimes_{U(\mathfrak{b})} (F \otimes \mathbb{C}_{\nu-\rho})$ and so inducing the flag B_i from $\mathfrak{b}$ up to $\mathfrak{g}$ we obtain a flag of $\mathfrak{g}$-modules: $F \otimes M(\nu) = A_r \supset \cdots \supset A_0 = 0$ with $A_i/A_{i-1} \simeq M(\mu_i)$, $1 \leq i \leq r$. The weights $\mu_i - \nu$ are the weights of F with multiplicity; and so, we check that $p_{\mu+\xi}M(\mu_i)$ equals zero unless $\mu_i = \nu'$. This gives $\varphi M(\nu) = M(\nu')$. The proof for ψ is similar.

Next we show $\varphi L(\nu) = L(\nu')$. Since φ is exact, $\varphi L(\nu)$ is a quotient of $\varphi M(\nu) = M(\nu')$. So we need only show that $\varphi L(\nu)$ is irreducible. The Verma modules admit invariant bilinear forms (cf. [**7**] where they are called contravariant) and these forms have the maximal proper submodule as radical. Since both F and $L(\nu)$ admit nondegenerate invariant forms so does $F \otimes L(\nu)$ and then $\varphi L(\nu)$. This implies that $\varphi L(\nu)$ is the irreducible quotient of $M(\nu')$. The proof for ψ is similar. This proves (3.8).

4. The Zuckerman derived functors. The link between the category $\mathcal{O}$ and the category of Harish-Chandra modules is given most satisfactorily by the use of techniques from homological algebra. The basic connection is given by the Zuckerman derived functors. In this section we define these functors and summarize their basic properties. Proofs of these results can be found in [**4**] and [**10**]. For the standard notions of derived functors, homotopy, and natural equivalence of functors, the reader should consult [**5**].

Let $\mathfrak{a}$ be a finite-dimensional Lie algebra over F and let $\mathfrak{b}$ be a subalgebra. Denote by $\mathcal{C}(\mathfrak{a}, \mathfrak{b})$ the category of all $\mathfrak{a}$-modules which as $\mathfrak{b}$-modules are completely reducible and $U(\mathfrak{b})$-finite. Let $\mathcal{A}(\mathfrak{a}, \mathfrak{b})$ be the subcategory of $\mathcal{C}(\mathfrak{a}, \mathfrak{b})$ whose objects A have finite multiplicity $\mathfrak{b}$-isotypic subspaces; i.e., for all irreducible $\mathfrak{b}$-modules F, $\mathrm{Hom}_{\mathfrak{b}}(F, A)$ is finite dimensional. For any $\mathfrak{a}$-module X, let $X[\mathfrak{b}]$ denote the sum of all irreducible finite-dimensional $\mathfrak{b}$-submodules of X. If $\mathfrak{b}$ is reductive in $\mathfrak{a}$ (i.e., $\mathfrak{a}$ is completely reducible as a $\mathfrak{b}$-module) then $X[\mathfrak{b}]$ is a $\mathfrak{a}$-submodule of X. For any module $W \in \mathcal{C}(\mathfrak{b}, \mathfrak{b})$, let $I(W) = \mathrm{Hom}_{\mathfrak{b}}(U(\mathfrak{a}), W)[\mathfrak{b}]$. If j is defined by $j(f) = f(1)$, then j induces a $\mathfrak{b}$-module projection from $I(W)$ onto W. In turn this projection gives a natural bijection

$$\mathrm{Hom}_{\mathfrak{a}}(V, I(W)) \xrightarrow{\sim} \mathrm{Hom}_{\mathfrak{b}}(V, W) \quad \text{for} \quad V \in \mathcal{C}(\mathfrak{a}, \mathfrak{b}).$$

This bijection implies that $I(W)$ is injective in $\mathcal{C}(\mathfrak{a}, \mathfrak{b})$. For $V \in \mathcal{C}(\mathfrak{a}, \mathfrak{b}), v \in V$, define $f_v \in I(V)$ by $f_v(x) = x \cdot v, x \in U(\mathfrak{a})$. Then $v \mapsto f_v$ is an injection. Clearly this implies that every object in $\mathcal{C}(\mathfrak{a}, \mathfrak{b})$ has an injective resolution.

We now fix our notation for the remainder of this section. Let $\mathfrak{g}$ be a finite-dimensional Lie algebra over F and $\mathfrak{t}$ and $\mathfrak{h}$ subalgebras with $\mathfrak{g} \supset \mathfrak{t} \supset \mathfrak{h}$. Assume $\mathfrak{t}$ is reductive and reductive in $\mathfrak{g}$. Assume $\mathfrak{h}$ is reductive and reductive in both $\mathfrak{t}$ and $\mathfrak{g}$. Let f be the forgetful functor $f\colon \mathcal{C}(\mathfrak{g}, \mathfrak{h}) \to \mathcal{C}(\mathfrak{t}, \mathfrak{h})$. For $W \in \mathcal{C}(\mathfrak{h}, \mathfrak{h})$, let $I(W) = \mathrm{Hom}_{\mathfrak{h}}(U(\mathfrak{t}), W)[\mathfrak{h}]$ and $I_{\mathfrak{g}}(W) = \mathrm{Hom}_{\mathfrak{h}}(U(\mathfrak{g}), W)[\mathfrak{h}]$. Every injective object in $\mathcal{C}(\mathfrak{g}, \mathfrak{h})$ is a summand of some $I_{\mathfrak{g}}(W)$. It is easy to show that $fI_{\mathfrak{g}}(W)$ is injective in $\mathcal{C}(\mathfrak{t}, \mathfrak{h})$; and so, f maps an injective resolution of $A \in \mathcal{C}(\mathfrak{g}, \mathfrak{h})$ to an injective resolution of fA.

For $A \in \mathcal{C}(\mathfrak{t}, \mathfrak{h})$ (resp. $\mathcal{C}(\mathfrak{g}, \mathfrak{h})$) define $\Gamma A = A[\mathfrak{t}]$ (resp. $\Gamma_{\mathfrak{g}} A = A[\mathfrak{t}]$). Then $\Gamma\colon \mathcal{C}(\mathfrak{t}, \mathfrak{h}) \to \mathcal{C}(\mathfrak{t}, \mathfrak{t})$ and $\Gamma_{\mathfrak{g}}\colon \mathcal{C}(\mathfrak{g}, \mathfrak{h}) \to \mathcal{C}(\mathfrak{g}, \mathfrak{t})$. Define the (right) derived functors of Γ and $\Gamma_{\mathfrak{g}}$ in the standard way. For $A \in \mathcal{C}(\mathfrak{t}, \mathfrak{h})$, let $0 \to A \to I^*$ be an injective resolution of A. Then delete A and apply Γ. The ith cohomology of the resulting complex is $\Gamma^i A$, the ith right derived functor of Γ applied to A. Since Γ is left exact, $\Gamma = \Gamma^0$.

PROPOSITION 4.1. *For all $i \in \mathbb{N}$, $f \circ \Gamma^i_{\mathfrak{g}}$ and $\Gamma^i \circ f$ are naturally equivalent functors from $C(\mathfrak{g}, \mathfrak{h})$ to $C(\mathfrak{t}, \mathfrak{t})$.*

To prove (4.1) let $A \in C(\mathfrak{g}, \mathfrak{h})$ and let $0 \to A \to I^*$ be an injective resolution of A. Now $0 \to fA \to fI^*$ is an injective resolution of fA in $C(\mathfrak{t}, \mathfrak{h})$ and $f\Gamma I^j = \Gamma f I^j$. So, $f\Gamma^i A = \Gamma^i fA$.

This result is essential for the theory. It asserts that the $\mathfrak{t}$-structure of $\Gamma^i_{\mathfrak{g}} A$ depends only on the $\mathfrak{t}$-structure of A. Moreover, there are important applications where computation of Γ^i in $C(\mathfrak{t}, \mathfrak{h})$ can be made. We now describe this basic computation which, in fact, is equivalent to the Bott–Borel–Weil theorem. We begin with a result regarding infinitesimal characters.

PROPOSITION 4.2. *Let χ be a character of $Z(\mathfrak{g})$ and let $X \in C(\mathfrak{g}, \mathfrak{h})$ with infinitesimal character χ. For any i, if $\Gamma^i_{\mathfrak{g}} X \neq 0$ then this $\mathfrak{g}$-module has infinitesimal character χ also.*

Proof. Let $0 \to X \to I^*$ be an injective resolution of X. Fix $z \in Z(\mathfrak{g})$ and let $u = z - \chi(z)$. Then $u \in Z(\mathfrak{g})$; and so, u acts on I^* as a map of chains of $\mathfrak{g}$-modules and induces the action on $\Gamma^i_{\mathfrak{g}} X$. However, since u acts by zero on X, the induced action is zero. So $\Gamma^i_{\mathfrak{g}} X$ has infinitesimal character χ. (The reader not familiar with the ingredients of this argument should consult [**5**, Theorem 4.1, p. 127]).

This argument actually proves somewhat more. It describes the action of the centralizer of $\mathfrak{t}$ in $U(\mathfrak{g})$ on $\Gamma^i_{\mathfrak{g}} X$ in terms of the action on X.

We now turn to the computation of $\Gamma^i X$ when X is a Verma module in $C(\mathfrak{t}, \mathfrak{h})$. Now assume $\mathfrak{h}$ is a Cartan subalgebra of $\mathfrak{t}$ and let notation be as in section two with $\mathfrak{g}$ replaced by $\mathfrak{t}$. Let $\ell(\cdot)$ denote the length function on the Weyl group.

THEOREM 4.3. *Let λ be dominant for $\Delta^+(\mathfrak{t}, \mathfrak{h})$ and let ω be an element of the Weyl group $\mathfrak{w}(\mathfrak{t}, \mathfrak{h})$. Then*

$$\Gamma^i M(\omega\lambda) = \begin{cases} L(\lambda) & \text{if } \lambda \text{ is integral and regular and} \\ & i = \dim(\mathfrak{t}/\mathfrak{h}) - \ell(\omega) \\ 0 & \text{otherwise.} \end{cases}$$

Sketch of proof. The finite-dimensional representations have $Z(\mathfrak{t})$-characters, χ_μ with μ regular and integral. So (4.2) applied with $\mathfrak{g} = \mathfrak{t}$ shows that $\Gamma^i M(\omega\lambda) = 0$ for all i unless λ is integral and regular. In $C(\mathfrak{t}, \mathfrak{h})$, tensoring by finite-dimensional $\mathfrak{t}$-modules maps injectives to injectives; and so, the Γ^i commute with tensoring by finite-dimensional $\mathfrak{t}$-modules. This fact and (4.2) imply that the Γ^i commute with the Zuckerman translation functors (cf. (3.8)). Using these translation functors and (3.8) one can easily reduce the proof of (4.3) to the case $\lambda = \rho$.

Now assume $\lambda = \rho$. We proceed by induction on $j = |\Delta^+| - \ell(\omega)$. If $j = 0$ then $\omega\rho = -\rho$. In this case one can verify the theorem for $\Gamma^i M(-\rho)$ directly by relating the Γ^i and Lie algebra cohomology for $\mathfrak{n}$. For details of this calculation see [**4**, Lemma 6.1]. If $j > 0$ then choose a simple root α so that $\ell(s_\alpha\omega) = \ell(\omega)+1$.

By applying the translation functor φ to a Verma module on the wall $\alpha \equiv 0$ and then translating to infinitesimal character χ_ρ we obtain a module E with $\Gamma^i E = 0$ for all i and a short exact sequence $0 \to M(\omega\rho) \to E \to M(s_\alpha\omega\rho) \to 0$. This short exact sequence induces a long exact sequence in cohomology. Since $\Gamma^i E = 0$ for all i, $\Gamma^i(s_\alpha\omega\rho) \cong \Gamma^{i+1}(\omega\rho)$. This isomorphism and the inductive hypothesis now prove (4.3) for ω. This completes the proof.

One of the deeper and especially useful properties of the derived functors is the following duality theorem. For an object X in $\mathcal{C}(\mathfrak{g}, \mathfrak{h})$ let $\tilde{X}$ be the $U(\mathfrak{h})$-locally finite vectors in the algebraic dual to X. If $X \in \mathcal{C}(\mathfrak{g}, \mathfrak{t})$, let $\tilde{\tilde{X}}$ be the $U(\mathfrak{t})$-locally finite vectors in the algebraic dual to X. Here the algebraic dual could be replaced by the contravariant dual used by Jantzen if applications require.

THEOREM 4.4. *For* $i \in \mathbb{N}$, $0 \leq i \leq m = \dim(\mathfrak{t}/\mathfrak{h})$, *the functors* $X \to \Gamma^i(\tilde{X})$ *and* $X \to (\Gamma^{m-i}(X))^{\approx}$ *are naturally equivalent functors from* $\mathcal{A}(\mathfrak{g}, \mathfrak{h})$ *to* $\mathcal{C}(\mathfrak{g}, \mathfrak{t})$.

Most applications of (4.4) deal with the middle dimension $s = \frac{1}{2}m$. Then if X admits a nondegenerate invariant form (i.e., if X and $\tilde{X}$ are isomorphic) then the same is true for the module $\Gamma^s(X)$.

BIBLIOGRAPHY

[1] Bernstein, I. N., Gel′fand, I. M., and Gel′fand, S. I., *Differential operators on the base affine space and a study of* $\mathfrak{g}$*-modules*, Lie Groups and Their Representations, Wiley, New York, 1975, 39–64.

[2] ——, *A certain category of* $\mathfrak{g}$*-modules*, Funkcional. Anal. i Priložen. (2), **10** (1976), 1–8.

[3] Dixmier, J., *Algèbres Enveloppantes*, Gauthier-Villars, Paris, 1974.

[4] Enright, T. J., and Wallach, N. R., *Notes on homological algebra and representations of Lie algebras*, Duke Math. J. **47** (1980), 1–15.

[5] Hilton, P. J., and Stammbach, U., *A Course in Homological Algebra*, Springer-Verlag, Berlin and New York, 1971.

[6] Humphreys, J., *Introduction to Lie Algebras and Representation Theory*, Springer-Verlag, Berlin and New York, 1972.

[7] Jantzen, J. C., *Moduln mit einem höchsten Gewicht*, Lecture Notes in Mathematics #750, Springer-Verlag, Berlin and New York, 1979.

[8] Kazhdan, D. and Lusztig, G., *Representations of Coxeter groups and Hecke algebras*, Inventiones Math. **53** (1979), 165–184.

[9] Varadarajan, V. S., *Lie Groups, Lie Algebras and Their Representations*, Prentice Hall, Englewood Cliffs, New Jersey, 1974.

[10] Vogan, Jr., D., *Representations of Real Reductive Lie Groups*, Birkhäuser, Boston-Basel-Stuttgart, 1981.

[11] Zuckerman, G., *Tensor products of infinite-dimensional and finite-dimensional representations of semisimple Lie groups*, Ann. of Math. **106** (1977), 295–308.

Primitive Ideals in the Enveloping Algebra of a Semisimple Lie Algebra

J. C. JANTZEN

1. Let $\mathfrak{g}$ be a finite-dimensional, semisimple complex Lie algebra. I want to give a report about the work done during the last years on the description of the set $\mathfrak{X}$ of primitive ideals in the enveloping algebra $U(\mathfrak{g})$ of $\mathfrak{g}$ and on the structure of the rings $U(\mathfrak{g})/I$ with $I \in \mathfrak{X}$. Complete proofs can be found in the original papers quoted and in [EA].

2. The algebra $U(\mathfrak{g})$ is Noetherian and has no zero-divisors. There is a natural filtration $U_0(\mathfrak{g}) \subset U_1(\mathfrak{g}) \subset U_2(\mathfrak{g}) \subset \cdots$ of $U(\mathfrak{g})$ and the associated graded algebra is isomorphic to the symmetric algebra $S(\mathfrak{g})$ on $\mathfrak{g}$ [10, 2.3]. For a finitely generated $U(\mathfrak{g})$-module we may choose a finite-dimensional subspace M_0 of M with $M = U(\mathfrak{g})M_0$ and can now define a filtration of M by $M_n = U_n(\mathfrak{g})M_0$. We get then a graded $S(\mathfrak{g})$-module $\operatorname{gr} M$ and its Hilbert polynomial giving the function $n \mapsto \dim(M_n)$ for large n. We may write the leading term of this polynomial in the form $e(M)n^{d(M)}/(d(M)!)$. Then the (natural) numbers $d(M)$ and $e(M)$ are independent of the choice of M_0. They are called the *Gel'fand–Kirillov dimension* of M and the *multiplicity* (or Bernstein degree) of M ([21], [14], [15], [23], [EA, Kap. 8]). Notions like critical or homogeneous will later on always be used with respect to this dimension d.

3. We shall always identify $(\mathfrak{g} \times \mathfrak{g})$-modules and $(U(\mathfrak{g}), U(\mathfrak{g}))$-bimodules. (On a bimodule X an element $(a, b) \in \mathfrak{g} \times \mathfrak{g}$ will operate via $(a, b)x = ax - xb$ for all $x \in X$.) Taking the usual structure as bimodule we may thus consider $U(\mathfrak{g})$ as $(\mathfrak{g} \times \mathfrak{g})$-module. The submodules are the (two-sided) ideals of $U(\mathfrak{g})$. We can identify the subalgebra $\mathfrak{k} = \{(a, a) | a \in \mathfrak{g}\}$ of $\mathfrak{g} \times \mathfrak{g}$ with $\mathfrak{g}$. The operation of $\mathfrak{g} \times \mathfrak{g}$ on $U(\mathfrak{g})$ restricted to $\mathfrak{k} \simeq \mathfrak{g}$ is the usual adjoint operation. Therefore all $U_n(\mathfrak{g})$ are stable under $U(\mathfrak{k})$, and $U(\mathfrak{g})$ is a locally finite $U(\mathfrak{k})$-module.

We call a $(\mathfrak{g} \times \mathfrak{g})$-module X a *Harish–Chandra* module if and only if (HC1) X is a locally finite $U(\mathfrak{k})$-module, and (HC2) for all finite-dimensional $\mathfrak{k}$-modules E we have $\dim \operatorname{Hom}_{\mathfrak{k}}(E, X) < \infty$. Condition (HC1) means that $\dim U(\mathfrak{k})x < \infty$ holds for all $x \in X$; thus X is a direct sum of finite-dimensional simple $\mathfrak{k}$-modules, as $\mathfrak{k}$ is semisimple. Condition (HC2) then means that each simple $\mathfrak{k}$-module is isomorphic only to a finite number of summands. Now $U(\mathfrak{g})$ satisfies (HC1) by the discussion above but not (HC2). Let $Z(\mathfrak{g})$ be the center of $U(\mathfrak{g})$. For any ideal I of $U(\mathfrak{g})$ the ring $U(\mathfrak{g})/I$ is a Harish–Chandra module if and only if $I \cap Z(\mathfrak{g})$ has finite codimension in $Z(\mathfrak{g})$. By a lemma of Quillen [10, 2.6.5] any simple $\mathfrak{g}$-module admits a central character. Hence any $I \in \mathfrak{X}$ intersects $Z(\mathfrak{g})$ in a maximal ideal of codimension 1. Therefore $U(\mathfrak{g})/I$ is a Harish–Chandra module.

4. It follows from Harish–Chandra's theory (cf. [10, 9.7.8], [5], [EA, 6.30]) that a Harish–Chandra module X has finite length if $Z(\mathfrak{g} \times \mathfrak{g}) \simeq Z(\mathfrak{g}) \otimes Z(\mathfrak{g})$ operates on X through a character. Hence $U(\mathfrak{g})/I$ has finite length for all $I \in \mathfrak{X}$. The definition of a prime ideal implies now ([11], [EA, 1.13]) that the socle $\operatorname{soc} U(\mathfrak{g})/I$ of $U(\mathfrak{g})/I$ is simple with

$$I = \operatorname{LAnn}(\operatorname{soc} U(\mathfrak{g})/I) = \operatorname{RAnn}(\operatorname{soc} U(\mathfrak{g})/I). \tag{1}$$

(For each $(U(\mathfrak{g}), U(\mathfrak{g}))$-bimodule X we denote its left annihilator in $U(\mathfrak{g})$ by $\operatorname{LAnn} X$ and its right annihilator by $\operatorname{RAnn} X$.)

We can turn a $\mathfrak{g}$-module E into a $(\mathfrak{g} \times \mathfrak{g})$-module E^ℓ making an element $(a, b) \in \mathfrak{g} \times \mathfrak{g}$ act as a. For each finitely generated Harish-Chandra module X we can find a finite-dimensional $\mathfrak{g}$-module E and exact sequences of $(\mathfrak{g} \times \mathfrak{g})$-modules ([24], [EA, 6.11]):

$$(U(\mathfrak{g})/\operatorname{RAnn} X) \otimes E^\ell \to X \to 0 \tag{2}$$

and

$$0 \to U(\mathfrak{g})/\operatorname{RAnn} X \to X \otimes (E^*)^\ell. \tag{3}$$

(Take a finite-dimensional subspace $E \subset X$ with $U(\mathfrak{k})E = E$ and $X = U(\mathfrak{g})EU(\mathfrak{g})$. Consider E as a $\mathfrak{g}$-module via $\mathfrak{g} \simeq \mathfrak{k}$ and take right multiplication as the map in (2), tensor (2) with E^*, and look at the identity in $\operatorname{End}(E) \simeq E \otimes E^*$.) There are similar maps for $\operatorname{LAnn} X$ instead of $\operatorname{RAnn} X$. From (2), (3) we can easily deduce

$$d(X) = d(U(\mathfrak{g})/\operatorname{RAnn} X) = d(U(\mathfrak{g})/\operatorname{LAnn} X) \tag{4}$$

([15], I, [EA, 10.3]). Observe that $d(X)$ is the Gel'fand–Kirillov dimension of X considered as $(\mathfrak{g} \times \mathfrak{g})$-module, but does not change if we regard X as a left $U(\mathfrak{g})$-module or a right $U(\mathfrak{g})$-module. A similar statement holds for $e(X)$.

For two finitely generated Harish–Chandra modules X, X' the inclusion $\operatorname{RAnn} X \subset \operatorname{RAnn} X'$ holds if and only if X' is a subquotient of $X \otimes E^\ell$ for some finite-dimensional $\mathfrak{g}$-module E ([24], [EA, 7.12]). Here one direction is obvious because of $\operatorname{RAnn} X = \operatorname{RAnn}(X \otimes E^\ell)$. If on the other hand $\operatorname{RAnn} X \subset \operatorname{RAnn} X'$ we combine the natural surjection $U(\mathfrak{g})/\operatorname{RAnn} X \to U(\mathfrak{g})/\operatorname{RAnn} X'$ with (2) for X' and (3) for X. By the last result and (1) we see how to determine $\mathfrak{X}$ from a thorough knowledge of the category of Harish–Chandra modules.

5. For two $\mathfrak{g}$-modules M, N the space $\mathrm{Hom}_{\mathbb{C}}(M,N)$ is a $(U(\mathfrak{g}),U(\mathfrak{g}))$-bimodule in a natural way. Therefore we can construct the sub-bimodule

$$\mathcal{L}(M,N) = \{\varphi \in \mathrm{Hom}_{\mathbb{C}}(M,N) \mid \dim U(\mathfrak{k})\varphi < \infty\}$$

satisfying (HC1). If we want to check (HC2) we have to look at

$$(5)\quad \begin{aligned}\mathrm{Hom}_{\mathfrak{k}}(E, \mathcal{L}(M,N)) &\simeq \mathrm{Hom}_{\mathfrak{k}}(E, \mathrm{Hom}_{\mathbb{C}}(M,N)) \\ &\simeq \mathrm{Hom}_{\mathfrak{g}}(M\otimes E, N) \simeq \mathrm{Hom}_{\mathfrak{g}}(M, N\otimes E^*)\end{aligned}$$

for each finite-dimensional $\mathfrak{g}$- (or $\mathfrak{k}$-) module E. Obviously $\mathcal{L}$ is a functor, left exact in both arguments. For $\varphi_1 \in \mathcal{L}(M_1,M_2)$ and $\varphi_2 \in \mathcal{L}(M_2,M_3)$ one has $\varphi_2\varphi_1 \in \mathcal{L}(M_1,M_3)$. Hence each $\mathcal{L}(M,M)$ is a ring in a natural way.

If M, N are $\mathfrak{g}$-modules of finite length, $\mathcal{L}(M,N)$ is a Harish–Chandra module of finite length ([22, III], [EA, 8.17]). Using the left exactness of $\mathcal{L}$ one has to prove this only for simple M and N. For them $Z(\mathfrak{g}\times\mathfrak{g})$ will operate through a character on $\mathcal{L}(M,N)$ so that we only have to check (HC2), cf. **3**. Now elementary properties of the Gel'fand–Kirillov dimension and of the multiplicity show $\mathrm{Hom}_{\mathfrak{g}}(M, N\otimes E^*) = 0$ for $d(M) \neq d(N)$ and $\dim \mathrm{Hom}_{\mathfrak{g}}(M, N\otimes E^*) \leq e(N\otimes E^*)/e(M) = e(N)\dim(E)/e(M) < \infty$ for $d(M) = d(N)$. Thus (HC2) follows from (5).

As a corollary we get that $\mathcal{L}(M,M)$ is a Noetherian ring for any $\mathfrak{g}$-module M of finite length, because $\mathcal{L}(M,M)$ is finitely generated as a left $U(\mathfrak{g})$-module. There is a natural embedding of $U(\mathfrak{g})/\mathrm{Ann}\, M$ into $\mathcal{L}(M,M)$ and one knows

$$(6)\qquad d(U(\mathfrak{g})/\mathrm{Ann}\, M) = d(\mathcal{L}(M,M)) \leq 2d(M)$$

([14], [20], [EA, 10.4(7)]). One expects equality to hold in (6) and it has been proved in many cases (cf. (7) below) but the general case is still open.

Kostant asked whether $U(\mathfrak{g})$ maps onto $\mathcal{L}(M,M)$ for simple M in the construction above. The answer (by Conze–Berline and Duflo) is "no"; there are counterexamples already for $\mathfrak{sl}_2$ ([19]). For critical (e.g., simple) M the rings $\mathcal{L}(M,M)$ and $U(\mathfrak{g})/\mathrm{Ann}\, M$ are prime and one has an inclusion of total quotient rings $Q(U(\mathfrak{g})/\mathrm{Ann}\, M) \subset Q(\mathcal{L}(M,M))$ with equality if and only if $e(U(\mathfrak{g})/\mathrm{Ann}\, M) = e(\mathcal{L}(M,M))$; cf. [16], [18], [EA, 12.1]. Then the Goldie rank $\mathrm{grk}\, U(\mathfrak{g})/\mathrm{Ann}\, M$ divides $\mathrm{grk}\, \mathcal{L}(M,M)$ and one would like to know the quotient as it is often easier to compute $\mathrm{grk}\, \mathcal{L}(M,M)$ than $\mathrm{grk}\, U(\mathfrak{g})/\mathrm{Ann}\, M$.

The ring $\mathcal{L}(M,M)$ may be prime even if $\mathrm{Ann}\, M$ is not a prime ideal. When Joseph and Small proved their additivity principle in [21] (cf. [22]) they wanted to apply it to this type of situation. In those cases where the principle is actually used in the theory of enveloping algebras one has better results than just the principle. Let us call (for the purpose of this talk only) a $\mathfrak{g}$-module M of finite length *wonderful*, if $\mathcal{L}(M,M)$ is prime and satisfies

$$\mathrm{grk}\, \mathcal{L}(M,M) = \sum_{L} [M : L]\, \mathrm{grk}\, \mathcal{L}(L,L),$$

where we sum over all $\mathfrak{g}$-modules L (up to isomorphism) with $d(M) = d(L)$ and denote by $[M : L]$ the multiplicity of L as a composition factor of M.

There are two important classes of wonderful modules. Let me mention one at once. Take a parabolic subalgebra $\mathfrak{p}$ of $\mathfrak{g}$ and a finite-dimensional simple $\mathfrak{p}$-module E. Construct the induced module $M_{\mathfrak{p}}(E) = U(\mathfrak{g}) \otimes_{U(\mathfrak{p})} E$. Then $M_{\mathfrak{p}}(E)$ is wonderful with $\operatorname{grk} \mathcal{L}(M_{\mathfrak{p}}(E), M_{\mathfrak{p}}(E)) = \dim E$ and the Goldie field of this ring is a Weyl skew field ([16], [12], [EA, 15.8/21/22]). In this case one has $d(\mathcal{L}(M_{\mathfrak{p}}(E), M_{\mathfrak{p}}(E))) = 2\dim(\mathfrak{g}/\mathfrak{p}) = 2d(M_{\mathfrak{p}}(E))$ by [6].

6. Let $\mathfrak{h}$ be a Cartan subalgebra and $\mathfrak{b} \supset \mathfrak{h}$ a Borel subalgebra of $\mathfrak{g}$. Because of $\mathfrak{b} = \mathfrak{h} \oplus [\mathfrak{b}, \mathfrak{b}]$ we can extend each linear form $\lambda \in \mathfrak{h}^*$ to a one-dimensional representation of $\mathfrak{b}$. Let $M(\lambda)$ be the induced $\mathfrak{g}$-module (the Verma module with highest weight λ). It admits a central character χ_λ and one has $\operatorname{Ann} M(\lambda) = U(\mathfrak{g})\ker(\chi_\lambda)$ by a theorem of Duflo (cf. [10, 8.4.3]). For any homomorphism $\chi\colon Z(\mathfrak{g}) \to \mathbb{C}$ one can find $\lambda \in \mathfrak{h}^*$ with $\chi = \chi_\lambda$ such that $M(\lambda)$ is simple. Therefore the $U(\mathfrak{g})\ker(\chi_\lambda)$ are the minimal primitive ideals in $U(\mathfrak{g})$. For all λ the natural embedding is in fact an isomorphism $U(\mathfrak{g})/U(\mathfrak{g})\ker(\chi_\lambda) \to \mathcal{L}(M(\lambda), M(\lambda))$; cf. [9], [16], [EA, 7.25].

For all $\lambda \in \mathfrak{h}^*$ the factor module $L(\lambda) = M(\lambda)/\operatorname{rad} M(\lambda)$ is simple, hence $I(\lambda) = \operatorname{Ann}_{U(\mathfrak{g})} L(\lambda)$ is primitive (i.e., in $\mathfrak{X}$). In this case one knows ([14], [EA, 10.8])

$$d(U(\mathfrak{g})/I(\lambda)) = 2d(L(\lambda)). \tag{7}$$

Now every simple Harish–Chandra module is isomorphic to some $\mathcal{L}(M(\lambda), L(\mu))$ for suitable $\lambda, \mu \in \mathfrak{h}^*$. (This follows from the more precise results in [5] or [17] (cf. [EA, 6.23/25]); the simple Harish–Chandra modules had originally been classified by Želobenko.) Note that not all $\mathcal{L}(M(\lambda), L(\mu))$ are simple and that a simple Harish–Chandra module may be written in more than one way as $\mathcal{L}(M(\lambda), L(\mu))$. For $\mathcal{L}(M(\lambda), L(\mu)) \neq 0$ one has obviously $\operatorname{LAnn} \mathcal{L}(M(\lambda), L(\mu)) = I(\mu)$. Therefore (1) implies

$$\mathfrak{X} = \{I(\lambda) | \lambda \in \mathfrak{h}^*\} \tag{8}$$

(cf. [11], [EA, 7.4]). For $\mathfrak{g} = \mathfrak{sl}_n$ one has for all $\lambda \in \mathfrak{h}^*$ an isomorphism of total quotient rings $Q(U(\mathfrak{g})/I(\lambda)) \xrightarrow{\sim} Q(\mathcal{L}(L(\lambda), L(\lambda)))$ and the Goldie field of $U(\mathfrak{g})/I(\lambda)$ is a Weyl skew field [16], [EA, 12.13/15.24]. For other $\mathfrak{g}$ the first statement is not true whereas the second one is conjectured to hold (Gel'fand–Kirillov conjecture).

Let W be the Weyl group of $(\mathfrak{g}, \mathfrak{h})$ and ρ the half sum of the positive roots. Write $w \cdot \lambda = w(\lambda + \rho) - \rho$ for all $w \in W$ and $\lambda \in \mathfrak{h}^*$. By Harish–Chandra one has $\chi_\lambda = \chi_\mu$ if and only if $\lambda \in W \cdot \mu$; cf. [10, 7.4]. Now $\mathfrak{X}$ is the disjoint union of the different $\mathfrak{X}_\lambda = \{I \in \mathfrak{X} \mid I \cap Z(\mathfrak{g}) = \ker(\chi_\lambda)\}$ and by (8) one has

$$\mathfrak{X}_\lambda = \{I(w \cdot \lambda) \mid w \in W\}. \tag{9}$$

The structure of this finite set $\mathfrak{X}_\lambda$ depends on integrality properties of λ only. Let us call λ *dominant* ($\lambda \in P(R)^{++}$) if $\dim L(\lambda) < \infty$ and *integral* ($\lambda \in P(R)$) if there is $w \in W$ such that $w(\lambda)$ is dominant. (If one identifies $\mathfrak{h}^*$ with $\mathbb{C}^\ell$ using a suitable basis, $P(R)$ is mapped into $\mathbb{Z}^\ell$ and $P(R)^{++}$ onto $\mathbb{N}^\ell$.) We shall look at $\mathfrak{X}_\lambda$ only for $\lambda \in P(R)$ where $\mathfrak{X}_\lambda$ has the most complicated structure, but where its description needs the least amount of notation.

7. Let us consider the category $\mathcal{O}$ of all $\mathfrak{g}$-modules M of finite length such that all composition factors of M are isomorphic to some $L(\mu)$ with $\mu \in P(R)$ and such that M is semisimple as an $\mathfrak{h}$-module. (This is only a subcategory of what is usually known as category $\mathcal{O}$; cf. [4], [EA, 4.3]). For M in $\mathcal{O}$ and $\mu \in P(R)$ we denote the multiplicity of $L(\mu)$ as a composition factor of M by $[M : L(\mu)]$. For all $\lambda \in -\rho + P(R)^{++}$ the Verma module $M(\lambda)$ is a projective object in $\mathcal{O}$, hence $M \mapsto \mathcal{L}(M(\lambda), M)$ is an exact functor from $\mathcal{O}$ to the category of Harish–Chandra modules.

For $\lambda \in P(R)^{++}$ this functor is an equivalence of categories between $\mathcal{O}$ and the category of all Harish–Chandra modules of finite length such that $Z(\mathfrak{g})$ operates on them from the right through χ_λ ([5], [EA, 6.27]). In this case the map $\mu \mapsto \mathcal{L}(M(\lambda), L(\mu))$ induces a bijection from $P(R)$ to the set of isomorphism classes of simple Harish–Chandra modules X with $\mathrm{RAnn}(X) \cap Z(\mathfrak{g}) = \ker(\chi_\lambda)$.

For a finite-dimensional $\mathfrak{g}$-module E and any M in $\mathcal{O}$ the module $M \otimes E$ is also in $\mathcal{O}$. For any λ one has an isomorphism $\mathcal{L}(M(\lambda), M \otimes E) \simeq \mathcal{L}(M(\lambda), M) \otimes E^\ell$. For $\lambda \in P(R)^{++}$ the composition factors of these tensor products have the form $\mathcal{L}(M(\lambda), L(\mu))$ with $\mu \in P(R)$, each one occurring with multiplicity $[M \otimes E \colon L(\mu)]$.

For all $\lambda, \mu \in P(R)^{++}$ and $w \in W$ the equality $I(w \cdot \mu) = \mathrm{LAnn}, \mathcal{L}(M(\lambda), L(w \cdot \mu))$ is obvious, whereas one has to do some work in order to prove that $\mathrm{RAnn}\, \mathcal{L}(M(\lambda), L(w \cdot \mu)) = I(w^{-1} \cdot \lambda)$; cf. [13], [EA, 7.9]. Combining these results with those from section **4** we can conclude now that $I(w_1 \cdot \lambda) \subset I(w_2 \cdot \lambda)$ holds for $w_1, w_2 \in W$ if and only if there is a finite-dimensional $\mathfrak{g}$-module E with $[L(w_1^{-1} \cdot \mu) \otimes E : L(w_2^{-1} \cdot \mu)] \neq 0$; cf. [24], [2], [EA, 7.13]. Thus the problem of describing $\mathfrak{X}_\lambda$ for $\lambda \in P(R)^{++}$ as an ordered set has been translated from a problem about Harish–Chandra modules into a problem about modules in the category $\mathcal{O}$.

8. The criterion for $I(w_1 \cdot \lambda) \subset I(w_2 \cdot \lambda)$ given at the end of the last section is obviously independent of λ. Thus for all $\lambda, \mu \in P(R)^{++}$ there is an isomorphism of ordered sets $T_\lambda^\mu \colon \mathfrak{X}_\lambda \to \mathfrak{X}_\mu$ with $T_\lambda^\mu I(w \cdot \lambda) = I(w \cdot \mu)$ for all $w \in W$; cf. [7], [EA, 5.8]. To argue in this way is not quite honest, however, as the computation of $\mathrm{RAnn}\, \mathcal{L}(M(\lambda), L(w \cdot \mu))$ above requires the transition from λ, μ to other dominant weights and the arguments used for these transitions give T_λ^μ directly. (One has also a refined theorem comparing $\mathfrak{X}_\lambda$ and $\mathfrak{X}_\mu$ for $\lambda \in P(R)^{++}$ and $\mu \in -\rho + P(R)^{++}$.)

Let me mention other properties of the translation map T_λ^μ for $\lambda, \mu \in P(R)^{++}$. For all $I \in \mathfrak{X}_\lambda$ the rings $U(\mathfrak{g})/I$ and $U(\mathfrak{g})/T_\lambda^\mu I$ have the same Gel'fand–Kirillov dimension ([7], [EA, 10.10]) and have isomorphic Goldie fields ([18], I, [EA,

12.4]). As the $(\mathfrak{g}\times\mathfrak{g})$-module $U(\mathfrak{g})/I$ has a simple socle (cf. **4**) there is a unique $w(I)\in W$ with $\operatorname{soc} U(\mathfrak{g})/I \simeq \mathcal{L}(M(\lambda), L(w(I)\cdot\lambda))$. Then one has $w(T_\lambda^\mu I) = w(I)$ for all $\mu \in P(R)^{++}$; cf. [18], I, [EA, 7.11(1)]. If we introduce functions q_w for all $w \in W$ and p_I for all $I \in \mathfrak{X}_\lambda$ by $q_w(\mu+\rho) = \operatorname{grk}\mathcal{L}(L(w\cdot\mu), L(w\cdot\mu))$ and $p_I(\mu+\rho) = \operatorname{grk} U(\mathfrak{g})/T_\lambda^\mu I$ for all $\mu \in P(R)^{++}$, then $p_I = q_{w(I)}$, as the total quotient rings $Q(U(\mathfrak{g})/I)$ and $Q(\mathcal{L}(L(w(I)\cdot\lambda), L(w(I)\cdot\lambda)))$ are naturally isomorphic ([18], I, [EA, 12.2]) without any restriction on the type of $\mathfrak{g}$ as in **6**.

9. In order to get better information about Goldie ranks we have to look again at the $\mathcal{L}(M(\lambda), M)$ with $\lambda \in P(R)^{++}$ and M in $\mathcal{O}$. The ring $\mathcal{L}(M, M)$ operates from the left on $\mathcal{L}(M(\lambda), M)$ and this operation commutes with the operation of $U(\mathfrak{g})$ from the right. One can even prove ([18], I, [EA, 6.37]):

$$\mathcal{L}(M, M) \simeq \operatorname{End} \mathcal{L}(M(\lambda), M)_{U(\mathfrak{g})}. \tag{10}$$

Using (10) one can prove ([EA, 12.3]): If M is homogeneous and if $I = \operatorname{RAnn}\mathcal{L}(M(\lambda), M)$ is primitive, then M is wonderful (in the sense of **5**) and the Goldie fields of $\mathcal{L}(M, M)$ and $\mathcal{L}(U(\mathfrak{g})/I)$ are isomorphic. This result can be applied to the $M_\mathfrak{p}(E)$ mentioned in **4** but also to $M = L(\mu)\otimes E$ for any $\mu \in P(R)$ and any finite-dimensional $\mathfrak{g}$-module E. Using the result in the second case one proves ([18], I, [EA, 12.6]) for all $\mu, \mu' \in P(R)$ and $w \in W$ the equation $|W| q_w(\mu) = \sum_{w'\in W} q_w(\mu + w'\mu')$, where q_w has been extended to $P(R)$ by "coherent continuation". Thus q_w has something like a "mean value property" from which it follows that q_w can be extended to a W-harmonic polynomial on $\mathfrak{h}^*$.

We may therefore consider $p_I = q_{w(I)}$ for $\lambda \in P(R)^{++}$ and $I \in \mathfrak{X}_\lambda$ as a polynomial on $\mathfrak{h}^*$, i.e., as an element of $S(\mathfrak{h})$. The group W operates on $S(\mathfrak{h})$ so that we can form the W-submodule $\Sigma(I) = \mathbb{C}[W]p_I$ generated by p_I. Then $\Sigma(I)$ is a simple $\mathbb{C}[W]$-module for all $I \in \mathfrak{X}_\lambda$. If $\Sigma(I)$ and $\Sigma(I')$ are isomorphic for $I, I' \in \mathfrak{X}_\lambda$ then $\Sigma(I) = \Sigma(I')$. The $p_{I'}$ with $I' \in \mathfrak{X}_\lambda$ and $\Sigma(I') = \Sigma(I)$ form a basis of $\Sigma(I)$; cf. [18], II, [EA, 14.16]. Let us call an irreducible representation of W *special* if it is isomorphic to $\Sigma(I)$ for some $I \in \mathfrak{X}_\lambda$. Then $|\mathfrak{X}_\lambda|$ is the sum of the degrees of the special representations of W.

For $\mathfrak{g} = \mathfrak{sl}_n$ all irreducible representations of W are special. This is not true for other simple Lie algebras. The set of special representations has been determined in all cases, cf. [1], [2].

10. The polynomials p_I can be computed up to a constant factor. The formulas are needed to prove the relations between these polynomials and the Weyl group representations described in the last section.

There are integers $a_{w,w'}$ such that we get an equation

$$[L(w\cdot\lambda)] = \sum_{w'\in W} a_{w,w'}[M(w'\cdot\lambda)] \tag{11}$$

(for all $\lambda \in P(R)^{++}$ and $w \in W$) in the Grothendieck group of $\mathcal{O}$, where we denote the class in this group of a module M by $[M]$. The coefficients $a_{w,w'}$ have been computed in [3] and [8], proving a conjecture by Kazhdan and Lusztig.

Now we can form the element $\mathbf{a}_w = \sum_{w' \in W} a_{w,w'}(w')^{-1}$ of $\mathbb{C}[W]$. Then we get $\mathbf{a}_w S^r(\mathfrak{h}) = 0$ for $r < m(w) = \dim(\mathfrak{g}/\mathfrak{b}) - d(L(w \cdot \lambda))$ where $\lambda \in P(R)^{++}$ and $\mathbf{a}_w S^{m(w)}(\mathfrak{h}) = \mathbb{C} p_{I(w \cdot \lambda)}$; cf. [18], II, [EA, 14.7]. In order to prove this one has to compare different measures for the size of a module. One gets as a by-product that for each $w \in W$ there is a positive integer $n(w)$ such that $n(w) \operatorname{rk} U(\mathfrak{g})/I(w \cdot \lambda)$ is the multiplicity of the $\mathfrak{g}$-module $L(w \cdot \lambda)$. The actual proof works with a different type of multiplicity. One makes $L(w \cdot \lambda)$ into a graded module $L(w \cdot \lambda) = \bigoplus_{n \geq 0} L(w \cdot \lambda)^n$ in a way compatible with a decomposition of $L(w \cdot \lambda)$ into its weight spaces with respect to $\mathfrak{h}$. Then the function $n \mapsto \sum_{i=0}^{n} \dim L(w \cdot \lambda)^i$ will be, for large n, a polynomial on residue classes of degree $d(L(w \cdot \lambda))$. The highest coefficient is constant and is the different type of "multiplicity" we need. It is not difficult to show that this "multiplicity" as a function of λ is in $\mathbf{a}_w S^{m(w)}(\mathfrak{h})$, but it requires a lot of work to prove that it is proportional to $p_{I(w \cdot \lambda)}$; cf. [18], II, [EA, Kap. 14].

Note added in proof. This survey was written early in 1983. Since then many new results have been proved in this theory that deserve to be added. However, let me make only one short remark: Nobody still expects to have equality in (6) for all M, since counterexamples have been constructed by J. T. Stafford, cf. Invent. Math. **79** (1985), 619–638.

REFERENCES

(In [EA] there is a more complete list of references.)

1. D. Barbasch and D. Vogan, *Primitive ideals and orbital integrals in complex classical groups*, Math. Ann. **259** (1982), 153–189.

2. D. Barbasch and D. Vogan, *Primitive ideals and orbital integrals in complex exceptional groups*, J. Algebra **80** (1983), 350–382.

3. A. Beilinson and J. Bernstein, *Localisation de $\mathfrak{g}$-modules*, C. R. Acad. Sci. Paris (I) **292** (1981), 15–18.

4. I. N. Bernštein, I. M. Gel'fand, and S. I. Gel'fand, *Category of $\mathfrak{g}$-modules*, Funct. Anal. Appl. **10** (1976), 87–92.

5. J. N. Bernstein and S. I. Gel'fand, *Tensor products of finite and infinite dimensional representations of semisimple Lie algebras*, Compositio Math. **41** (1980), 245–285.

6. W. Borho, *Berechnung der Gel'fand-Kirillov-Dimension bei induzierten Darstellungen*, Math. Ann. **225** (1977), 177–194.

7. W. Borho and J. C. Jantzen, *Über primitive Ideale in der Einhüllenden einer halbeinfachen Lie-Algebra*, Invent. Math. **39** (1977), 1–53.

8. J. L. Brylinski and M. Kashiwara, *Kazhdan-Lusztig conjecture and holonomic systems*, Invent. Math. **64** (1981), 387–410.

9. N. Conze, *Algèbres d'opérateurs différentiels et quotients des algèbres enveloppantes*, Bull. Soc. Math. France **102** (1974), 379–415.

10. J. Dixmier, *Algèbres Enveloppantes*, Paris-Bruxelles-Montréal, 1974.

11. M. Duflo, *Sur la classification des idéaux primitifs dans l'algèbre enveloppante d'une algèbre de Lie semi-simple*, Ann. of Math. **105** (1977), 107–120.

12. A. Joseph, *On the Gel'fand-Kirillov conjecture for induced ideals in the semisimple case*, Bull. Soc. Math. France **107** (1979), 139–159.

13. A. Joseph, *On the annihilators of the simple subquotients of the principal series*, Ann. Scient. Éc. Norm Sup. (4) **10** (1977), 419–440.

14. A. Joseph, *Gel'fand-Kirillov dimension for the annihilators of simple quotients of Verma modules*, J. London Math. Soc. (2) **18** (1978), 50–60.

15. A. Joseph, *Towards the Jantzen conjecture I/II/III*, Compositio Math. **40** (1980), 35-67/**40** (1980), 69-78/**41** (1981), 23–30.

16. A. Joseph, *Kostant's problem, Goldie rank and the Gel'fand-Kirillov conjecture*, Invent. Math. **56** (1980), 191–213.

17. A. Joseph, *Dixmier's problem for Verma and principal series submodules*, J. London Math. Soc. (2) **20** (1979), 193–204.

18. A. Joseph, *Goldie rank in the enveloping algebra of a semisimple Lie algebra I/II/III*, J. Algebra **65** (1980), 269–283/**65** (1980), 284–306/**73** (1981), 295–326.

19. A. Joseph, *Kostant's problem and Goldie rank*, pp. 249–266 in *Non-Commutative Harmonic Analysis and Lie Groups*, Lecture Notes in Mathematics 880, Berlin-Heidelberg-New York, 1981.

20. A. Joseph, *Application de la théorie des anneaux aux algèbres enveloppantes*, Lecture Notes, Paris, 1981.

21. A. Joseph and L. W. Small, *An additivity principle for Goldie rank*, Israel J. Math. **31** (1978), 105–114.

22. J. T. Stafford, *The Goldie rank of a module*, this volume.

23. D. A. Vogan, *Gel'fand-Kirillov dimension for Harish-Chandra modules*, Invent. Math. **48** (1978), 75–98.

24. D. A. Vogan, *Ordering of the primitive spectrum of a semisimple Lie algebra*, Math. Ann. **248** (1980), 195–203.

[EA] J. C. Jantzen, *Einhüllende Algebren halbeinfacher Lie-Algebren*, Ergebnisse der Mathematik, 3. Folge, Band 3, Berlin-Heidelberg-New York-Tokyo, 1983.

Primitive Ideals in Enveloping Algebras (General Case)

R. RENTSCHLER

0. Introduction. Let $\mathfrak{g}$ be a finite-dimensional Lie algebra over an algebraically closed field k of characteristic zero. Let $U(\mathfrak{g})$ be the enveloping algebra of $\mathfrak{g}$.

There is now a complete classification of the primitive ideals of $U(\mathfrak{g})$ available by a reduction to the semisimple case and by using the two Duflo parameters (f, ξ) in the case where $\mathfrak{g}$ is algebraic.

The reduction uses Mackey theory as developed by Duflo, the Dixmier map in the nilpotent case, and the Duflo splitting. The Duflo splitting uses in an essential way the Dixmier map in the solvable case. It is also important to know that primitive ideals coincide with rational ideals. The classification map (using forms of unipotent type) and its surjectivity was established by Duflo in 1981; the injectivity (modulo conjugation) has been known since 1983. Until now it is not possible to determine the Goldie rank if one knows the Goldie rank of the second Duflo parameter. (Does one have equality?)

1. Primitive and rational ideals. A primitive ideal is by definition the annihilator of an irreducible module; hence it is prime. From the definition one should distinguish the left and the right side. We will take annihilators of irreducible left modules. In enveloping algebras this distinction between left and right will disappear.

If U is a prime Goldie ring, we denote by $Q(U)$ its total ring of fractions.

1.1. DEFINITION. A prime ideal I of a (left or right) Noetherian k-algebra U is called *rational* if the center of $Q(U/I)$ is the base field.

We shall use the following notation (U Noetherian):

$$\begin{aligned}
\operatorname{Spec} U &= \text{the space of prime ideals of } U,\\
\operatorname{Prim} U &= \text{the space of (left) primitive ideals of } U,\\
\operatorname{Rat} U &= \text{the space of rational ideals of } U,\\
\operatorname{Max} U &= \text{the space of maximal ideals of } U.
\end{aligned}$$

On all four spaces, we have the Jacobson topology. For example, if X is a subset of $\operatorname{Spec} U$, then a prime ideal I is in the closure $\overline{X}$ of X if and only if $I \supseteq \bigcap_{\mathfrak{p}\in X} \mathfrak{p}$. By $\hat{I}$ we denote the intersection of all prime ideals of U which contain I strictly.

1.2. For the rest of this paragraph, let $\mathfrak{g}$ be a Lie algebra over k and $U = U(\mathfrak{g})$ its enveloping algebra. Then $\mathfrak{g}$ operates by the adjoint action ad(?) by derivations on U. The group of automorphisms of $\mathfrak{g}$ acts on $U(\mathfrak{g})$ and by the contragredient operation on the dual space $\mathfrak{g}^*$ of $\mathfrak{g}$. We denote by Γ $(= \Gamma_{\mathfrak{g}})$ the adjoint algebraic group of $\mathfrak{g}$, i.e., the smallest algebraic subgroup of the group of automorphisms of $\mathfrak{g}$ such that its Lie algebra contains $\operatorname{ad}(\mathfrak{g})$.

By $A_n(k)$ we denote the nth Weyl algebra $k[x_1, \ldots, x_n, \partial/\partial x_1, \ldots, \partial/\partial x_n]$ over k (cf. [**Di**, 5, IV.6.3]). We recall that $A_n(k)$ is a simple algebra with center k and that all derivations of $A_n(k)$ are inner. We give now some particular information.

1.3. PROPOSITION (cf. [**Di**, 5, 3.7]). *If $\mathfrak{g}$ is solvable, then any prime ideal of $U(\mathfrak{g})$ is completely prime.*

1.4. PROPOSITION (cf. [**Di**, 5, 6.2] and [**M-R**, 3, III.3] for (ii)). *Let $\mathfrak{g}$ be nilpotent. Then we have the following.*

i) $\operatorname{Rat} U(\mathfrak{g}) = \operatorname{Max} U(\mathfrak{g}) = \operatorname{Prim} U(\mathfrak{g})$.

ii) *If M is a maximal ideal of $U(\mathfrak{g})$, then $U(\mathfrak{g})/M$ is isomorphic to a Weyl algebra $A_n(k) = k[x_1, \ldots, x_n, \partial/\partial x_1, \ldots, \partial/\partial x_n]$.*

iii) *In the situation of* (ii), *the isomorphism can be chosen such that the preimage of $k[x_1, \ldots, x_n]$ is $\operatorname{ad}(\mathfrak{g})$-stable.*

1.5. PROPOSITION (cf. [**Di**, 5, 3.1] and [**Du**, 2] for ii). *Let $U(\mathfrak{g})$ be an enveloping algebra over k. Then*

i) $\operatorname{Prim} U(\mathfrak{g}) \subseteq \operatorname{Rat} U(\mathfrak{g})$ ("*Quillen's Lemma*");

ii) *any prime ideal of $U(\mathfrak{g})$ is an intersection of primitive ideals* ("*$U(\mathfrak{g})$ is a Jacobson ring*");

iii) *if I is a prime ideal of $U(\mathfrak{g})$ such that $I \neq \hat{I}$, then I is primitive.*

If $\mathfrak{g}$ is solvable and if $I \in \operatorname{Rat} U(\mathfrak{g})$, then it is easy to see that there exists a $\mathfrak{g}$-semi-invariant element of $U(\mathfrak{g})/I$ contained in any prime ideal of $U(\mathfrak{g})/I$; hence $\hat{I} \neq I$. We get the following.

1.6. PROPOSITION. *If $\mathfrak{g}$ is a solvable Lie algebra and if $I \in \operatorname{Spec} U(\mathfrak{g})$, then the following properties are equivalent*:

i) *I is rational*;

ii) *I is primitive*;

iii) $I \neq \hat{I}$.

The same result is true for $\mathfrak{g}$ semisimple.

1.7. THEOREM. *The equivalences of Proposition* 1.4 *are also true for $\mathfrak{g}$ semisimple.*

This theorem is due to Dixmier who shows first that there are only finitely many prime ideals of $U(\mathfrak{g})$ ($\mathfrak{g}$ semisimple) containing a given maximal ideal M of the center of $U(\mathfrak{g})$. Later on, Duflo [**Du**, 3] showed the stronger result that $U(\mathfrak{g})/U(\mathfrak{g})M$ is a Harish–Chandra module (for $(\mathfrak{g}, \mathfrak{g}\times\mathfrak{g})$ with suitable actions) implying that $U(\mathfrak{g})/U(\mathfrak{g})M$ is of finite length as a $\mathfrak{g}$-bimodule (see also [**Ja**, Ch. 6–7]).

This characterization result for primitive ideals still remains true in the general case. There are the following two results.

1.8. THEOREM. *Let k be nondenumerable. Let I be a prime ideal of an enveloping algebra $U(\mathfrak{g})$. Then*

i) *I is primitive iff I is rational* [**Di**, 6];

ii) *I is primitive iff $I \neq \hat{I}$* [**M**].

1.9. COROLLARY (Dixmier–Moeglin equivalence). *Let $\mathfrak{g}$ be a Lie algebra over k. Then for a prime ideal I of $U(\mathfrak{g})$ the following properties are equivalent:*

i) *I is rational;*

ii) *I is primitive;*

iii) *$I \neq \hat{I}$.*

It remains only to point out that (i) $\Rightarrow$ (iii). Let k' be an algebraically closed extension field of k which is nondenumerable. Then $I\otimes k'$ is still rational; hence $\widehat{I\otimes k'} \neq I\otimes k'$. But this implies $\hat{I}\neq I$.

2. Orbits of an algebraic group. Let G be a linear algebraic k-group and U a Noetherian k-algebra on which G operates rationally by automorphisms (for example, if U is the enveloping algebra of an ideal of the Lie algebra of G). If J is a rational ideal of U and if $I = \bigcap_{\gamma\in G}\gamma J$, then it is not difficult to see that the G-invariants of the center of $Q(U/I)$ are reduced to k. We have:

2.1. THEOREM. *If ω_1 and ω_2 are two G-orbits in $\operatorname{Rat} U$ whose closures coincide, then we have $\omega_1=\omega_2$. Moreover, if I is a G-stable semiprime ideal of U such that the G-invariants of the center of $Q(U/I)$ are reduced to k, then there is a rational ideal J of U such that $I = \bigcap_{\gamma\in G}\gamma J$.*

For the proof, see [**M-R**, 1, 2.12, 3.8, and 3.11] and [**M-R**, 4].

The existence result ("existence of generic ideals"), first proven for enveloping algebras, is now available in general (for the origins of this problem, see [**Di**, 3]).

Now let $\mathfrak{g}$ be a Lie algebra over our base field and let P be a primitive (= rational) ideal of $U(\mathfrak{g})$. Let $\mathfrak{k}$ be an ideal of $\mathfrak{g}$ and Γ the adjoint algebraic group.

2.2. COROLLARY. *There exists a unique Γ-orbit ω in $\operatorname{Rat} U(\mathfrak{k})$ $(=\operatorname{Prim} U(\mathfrak{k}))$ (called the support of P with respect to $\mathfrak{k}$) such that*

$$P\cap U(\mathfrak{k}) = \bigcap_{\gamma\in\Gamma}\gamma J \quad \text{for any } J\in\omega.$$

PROOF. Let $I = P\cap U(\mathfrak{k})$. Then the Γ-invariants of $Q(U(\mathfrak{k})/I)$ are reduced to $\mathfrak{k}$. Theorem 2.1 gives the existence and the uniqueness of ω.

We would also like to mention

2.3. PROPOSITION. *If H is a linear algebraic group operating rationally by automorphisms on an enveloping algebra $U(\mathfrak{g})$ and if ω is a H-orbit in* $\operatorname{Rat} U(\mathfrak{g})$ $(= \operatorname{Prim} U(\mathfrak{g}))$, *then ω is open in its closure.*

See [**M-R**, 1, 3.10]. This property of orbits is due to the fact that $I \neq \hat{I}$ holds for rational ideals. For more details see also [**M-R**, 4].

3. The Dixmier map in the nilpotent case. Let $\mathfrak{g}$ be a nilpotent Lie algebra and let G be the unique unipotent linear algebraic k-group such that $\mathfrak{g}$ is its Lie algebra, Lie G. Let $\mathfrak{g}^*$ denote the dual space of $\mathfrak{g}$. We recall that any automorphism of $\mathfrak{g}$ extends to $U(\mathfrak{g})$ and operates in the contragredient way on $\mathfrak{g}^*$.

3.1. THEOREM ($\mathfrak{g}$ nilpotent). *There is a unique collection of maps*

$$I_{\mathfrak{h}}: \mathfrak{h}^* \to \operatorname{Prim} U(\mathfrak{h})$$

where $\mathfrak{h} = \operatorname{Lie} H$, H running through all algebraic subgroups of G such that

0) *$I_{\mathfrak{h}}$ is H-equivariant;*

i) *$I_{\mathfrak{h}}(f) = \operatorname{Ker}(f^U)$ if $f([\mathfrak{h},\mathfrak{h}]) = 0$ (where f^U is the homomorphism from $U(\mathfrak{h})$ to k, extending f);*

ii) *$I_{\mathfrak{h}}(f) \cap U(\mathfrak{h}_1) = \bigcap_{\gamma \in H} \gamma I_{\mathfrak{h}_1}(f_{|h_1})$ whenever $\mathfrak{h}$ is a subalgebra of $\mathfrak{g}$ and $\mathfrak{h}_1$ is an ideal of codimension 1 of $\mathfrak{h}$.*

Before discussing this theorem and indicating its proof we mention the following.

3.2. REMARK. The maps $I_{\mathfrak{h}}$ are necessarily injective modulo H-conjugation.

PROOF (by induction on $\dim \mathfrak{g}$). Let $f, f' \in \mathfrak{g}^*$ and $I_{\mathfrak{g}}(f) = I_{\mathfrak{g}}(f')$. We may suppose that $\mathfrak{g}(f)$ $(= \{x \in \mathfrak{g} | f([x, \mathfrak{g}]) = 0\})$ is not $\mathfrak{g}$. Let $\mathfrak{h}$ be an ideal of $\mathfrak{g}$ of codimension 1 containing $\mathfrak{g}(f)$. Then we have $\mathfrak{h}(f_{|\mathfrak{h}}) \not\subset \mathfrak{g}(f)$ and $H(f_{|\mathfrak{h}})f = f + \mathfrak{h}^{\perp}$ where $H(?)$ denotes the stabilizer in the group H and $\mathfrak{h}^{\perp}$ the orthogonal of $\mathfrak{h}$ in $\mathfrak{g}^*$. By (ii) the H-orbits of $I_{\mathfrak{h}}(f_{|\mathfrak{h}})$ and of $I_{\mathfrak{h}}(f'_{|\mathfrak{h}})$ have the same closure in $\operatorname{Rat} U(\mathfrak{g}) = \operatorname{Prim} U(\mathfrak{g}) = \operatorname{Max} U(\mathfrak{g})$. Hence by 2.1 we can assume that $f'_{|\mathfrak{h}} = f_{|\mathfrak{h}}$. (One could also show by other means that these orbits are closed.) But then $f' \in Gf$.

The existence of the maps $I_{\mathfrak{h}}$ will be given in the next paragraph (even for $\mathfrak{g}$ solvable). Here we can give the proof of the uniqueness by induction on $\dim \mathfrak{g}$.

Hence let us suppose we have two maps $I_{\mathfrak{g}}$, $I'_{\mathfrak{g}}$ for $\mathfrak{g}$ and let $f \in \mathfrak{g}^*$, $\mathfrak{g}(f) \neq \mathfrak{g}$. Let again $\mathfrak{h}$ be an ideal of codimension 1 containing $\mathfrak{g}(f)$. In this situation G cannot stabilize $H \cdot f_{|\mathfrak{h}}$. Now, by the injectivity of $I_{\mathfrak{h}}$ (modulo H), the prime ideal $I := \bigcap_{\gamma \in G} \gamma I_{\mathfrak{h}}(f_{|\mathfrak{h}})$ is not primitive and the operation of $\mathfrak{g}$ on $Q(U(\mathfrak{g})/I)$ cannot be inner. However, this implies that $U(\mathfrak{g})I$ is a primitive (= maximal) ideal of $U(\mathfrak{g})$ and hence $I_{\mathfrak{g}}(f) = U(\mathfrak{g})I = I'_{\mathfrak{g}}(f)$.

If $\lambda \in \mathfrak{g}^*$ such that $\lambda([\mathfrak{g},\mathfrak{g}]) = 0$, we define by τ_λ the automorphism of $U(\mathfrak{g})$ given by $x \mapsto x + \lambda(x)$ for $x \in \mathfrak{g}$. If $\mathfrak{h}$ is a subalgebra, we denote by $\tau_{\mathfrak{g}/\mathfrak{h}}$ the automorphism τ_θ of $U(\mathfrak{h})$ with $\theta(x) = -\frac{1}{2}\operatorname{trace}\operatorname{ad}_{\mathfrak{g}/\mathfrak{h}}(x) \qquad (x \in \mathfrak{h})$.

3.3. COROLLARY.
i) $I_\mathfrak{g}$ *commutes with any automorphism of* $\mathfrak{g}$.
ii) *If* $\lambda \in \mathfrak{g}^*$ *with* $\lambda([\mathfrak{g},\mathfrak{g}]) = 0$, *then we have*

$$I_\mathfrak{g}(f+\lambda) = \tau_{-\lambda} I_\mathfrak{g}(f).$$

PROOF. (i) is immediate from the unicity statement. Also (ii) is a consequence of the unicity statement by regarding the family

$$I'_\mathfrak{h}(f) := \tau_{-\lambda}(I_\mathfrak{h}(f - \lambda_{|\mathfrak{h}})), \quad f \in \mathfrak{h}^*.$$

4. Isotropy, coisotropy, polarization, induction.

4.1. Let $\mathfrak{g}$ be a Lie algebra over the base field k and $f \in \mathfrak{g}^*$. Then $f([x,y])$ is an alternating bilinear form on $\mathfrak{g}$ whose kernel is $\mathfrak{g}(f)$. If $\mathfrak{a} \subset \mathfrak{g}$ is a linear subspace, then we denote by $\mathfrak{a}^{f\perp}$ its orthogonal with respect to this form, i.e., $\mathfrak{a}^{f\perp} := \{x \in \mathfrak{g} | f([x,\mathfrak{a}]) = 0\}$. Hence we have $(\mathfrak{a}^{f\perp})^{f\perp} = \mathfrak{a} + \mathfrak{g}(f)$.

4.2. DEFINITION. We say that $\mathfrak{a}$ is *f-isotropic* if $f([\mathfrak{a},\mathfrak{a}]) = 0$. We say that $\mathfrak{a}$ is *f-coisotropic* if $\mathfrak{a}^{f\perp} \subseteq \mathfrak{a}$. We say that $\mathfrak{a}$ is a *polarization* for f if $\mathfrak{a}$ is a subalgebra and if $\mathfrak{a}$ is f-isotropic and f-coisotropic.

4.3. If $\mathfrak{h}$ is a subalgebra of $\mathfrak{g}$, then we denote by $\operatorname{ind}(?;\mathfrak{h}\uparrow\mathfrak{g})$ the induction for representations of $\mathfrak{h}$ and by Ind the induction for ideals of $U(\mathfrak{h})$. We use $\operatorname{ind}^\sim$ and $\operatorname{Ind}^\sim$ for the twisted inductions (cf. [**Di**, 5, 5.1 and 5.2]).

Hence if I is a two-sided ideal of $U(\mathfrak{h})$, then $\operatorname{Ind}^\sim(I;\mathfrak{h}\uparrow\mathfrak{g}) = \operatorname{Ind}(\tau_{\mathfrak{g}/\mathfrak{h}}I;\mathfrak{h}\uparrow\mathfrak{g})$ where $\tau_{\mathfrak{g}/\mathfrak{h}}$ is as above and $\operatorname{Ind}(I;\mathfrak{h}\uparrow\mathfrak{g}) = \bigcap_{\gamma\in\Gamma_\mathfrak{g}} \gamma(U(\mathfrak{g})I) =$ the largest two-sided ideal of $U(\mathfrak{g})$ contained in $U(\mathfrak{g})I$. For the relation between "left"-induction (this one) and "right"-induction, we refer to [**Du**, 6]. ($\Gamma_\mathfrak{g}$ = adjoint algebraic group.)

5. The Dixmier map in the solvable case.

5.1. NOTATIONS. Let $\mathfrak{g}$ be a Lie algebra over k and let Γ be its adjoint algebraic group. From 5.2 on until the end of this paragraph, we suppose $\mathfrak{g}$ to be solvable.

If $f \in \mathfrak{g}^*$ and if $\mathfrak{h}$ is any subalgebra of $\mathfrak{g}$ such that $f([\mathfrak{h},\mathfrak{h}]) = 0$, then we denote by $J(f,\mathfrak{h})$ the kernel in $U(\mathfrak{g})$ of the representation $\operatorname{ind}^\sim(f_{|\mathfrak{h}};\mathfrak{h}\uparrow\mathfrak{g})$, i.e.,

$$J(f,\mathfrak{h}) = \bigcap_{\gamma\in\Gamma} \gamma(U(\mathfrak{g})\tau_{-\theta}(\operatorname{Ker} f^U_{|\mathfrak{h}}))$$

where

$\theta(x) := \frac{1}{2}\operatorname{trace}(\operatorname{ad}_{\mathfrak{g}/\mathfrak{h}} x)$ for $x \in \mathfrak{h}$,

$f^U_{|\mathfrak{h}}$:= the extension of $f_{|\mathfrak{h}}$ to a homomorphism from $U(\mathfrak{h})$ in k.

(We have $\theta = 0$ if $\mathfrak{g}$ is nilpotent.)

5.2. PROPOSITION (Dixmier–Vergne). *If $\mathfrak{g}$ is solvable, then any $f \in \mathfrak{g}^*$ has a polarization.*

We give only the Vergne construction (cf. [**V**]). Let $n = \dim \mathfrak{g}$. If $S = (\mathfrak{g}_n, \mathfrak{g}_{n-1}, \ldots, \mathfrak{g}_0)$ is a sequence of ideals of $\mathfrak{g}$ with $\dim \mathfrak{g}_i = i$ and if f_i denotes the restriction of f to $\mathfrak{g}_i$, then $\sum_{i=1}^n \mathfrak{g}_i(f_i)$ is a polarization of f (called a Vergne polarization).

The following theorem allows us to define the Dixmier map in the solvable case.

5.3. THEOREM (Dixmier) (cf. [**Di**, 1] or [**B-G-R**, §10] or [**Di**, 5, Ch. 6]). *Let $\mathfrak{g}$ be a solvable Lie algebra and let $f \in \mathfrak{g}^*$. If $\mathfrak{h}_1$ and $\mathfrak{h}_2$ are two polarizations of f, then we have $J(f, \mathfrak{h}_1) = J(f, \mathfrak{h}_2)$.*

5.4. DEFINITION ($\mathfrak{g}$ solvable). If $f \in \mathfrak{g}^*$, we define $J(f) := J(f, \mathfrak{h})$ where $\mathfrak{h}$ is any polarization of f (sometimes we write also $J_{\mathfrak{g}}(f)$). If $\mathfrak{g}$ is nilpotent, we shall usually write $I_{\mathfrak{g}}$ (since $J_{\mathfrak{g}}$ will fulfill the properties required in §3 as we will mention in 5.10).

5.5. PROPOSITION (cf. [**Di**, 1] and [**Di**, 5, 6.1]). *If $f \in \mathfrak{g}^*$, $\mathfrak{g}$ solvable, then $J(f)$ is primitive.*

Indeed, one can show that $\operatorname{ind}^\sim(f_{|\mathfrak{h}}; \mathfrak{h} \uparrow \mathfrak{g})$ is irreducible if $\mathfrak{h}$ is a Vergne polarization of f.

5.6. For a recent study of properties of induced representations in the solvable case, see [**T**, 2].

5.7. REMARK. Let $\mathfrak{g}$ be a solvable Lie algebra.

i) The Dixmier map $J_{\mathfrak{g}}: \mathfrak{g}^* \to \operatorname{Prim} U(\mathfrak{g})$ is compatible with automorphisms a of $\mathfrak{g}$ where $af = f \circ a^{-1}$ for $f \in \mathfrak{g}^*$.

ii) If $f \in \mathfrak{g}^*$ and if $\mathfrak{h}$ is an ideal of $\mathfrak{g}$ contained in $\mathfrak{g}(f)$, then we have $x - f(x) \in J(f)$ for all $x \in \mathfrak{h}$.

These two properties are direct results of the definition of $J_{\mathfrak{g}}$.

5.8. THEOREM ($\mathfrak{g}$ solvable). *The Dixmier map $J_{\mathfrak{g}}: \mathfrak{g}^* \to \operatorname{Prim} U(\mathfrak{g})$ is surjective, continuous, and injective modulo Γ-conjugation.*

For the details (history and proof) we refer to [**B-G-R**] and to [**Di**, 5, Ch. 6]. The injectivity was first proved in [**R**], the continuity in [**C-Du**].

5.9. PROPOSITION ($\mathfrak{g}$ solvable). *If $f \in \mathfrak{g}^*$ and if $\mathfrak{h} \subseteq \mathfrak{g}$ is a subalgebra, then one has*

$$J_{\mathfrak{g}}(f) \cap U(\mathfrak{h}) = \bigcap_{\gamma \in \Gamma} J_{\mathfrak{h}}((\gamma f)_{|\mathfrak{h}}).$$

For the proof we refer to [**P**].

5.10. COROLLARY. *$J_{\mathfrak{g}}$ satisfies the conditions* 0), i), ii) *of Theorem* 3.1. *Hence we have the existence and uniqueness of $I_{\mathfrak{g}} = J_{\mathfrak{g}}$ in the nilpotent case. (This terminates the proof of Thm.* 3.1.)

5.11. REMARK ($\mathfrak{g}$ solvable). Let $\overline{J}_{\mathfrak{g}}: \mathfrak{g}^*/\Gamma \to \operatorname{Prim} U(\mathfrak{g})$ denote the factored Dixmier map. It is in general unknown whether $\overline{J}_{\mathfrak{g}}$ is bicontinuous. For some positive results on this question, see [**T**, 1].

In the nilpotent case, this question has a positive answer.

5.12. PROPOSITION (Berline). *If $\mathfrak{g}$ is nilpotent, then the factored Dixmier map $\mathfrak{g}^*/\Gamma \to \operatorname{Prim} U(\mathfrak{g}) = \operatorname{Max} U(\mathfrak{g})$ is bicontinuous.*

For the proof we refer to [**C**, 2].

5.13. The notation $I_{\mathfrak{g}}$ for the case where $\mathfrak{g}$ is nilpotent is due to the fact that there are two generalizations.

One generalization is the Dixmier map in the solvable case (and further the Dixmier–Duflo map in the general case, see §15). The other generalization will be the Duflo construction (notation $I_{\mathfrak{g}}(f, \xi)$) in the general (algebraic) case (§11).

We have the easy uniqueness result (§3) in the nilpotent case; we do not know whether a reasonable uniqueness result can be obtained in the solvable case.

6. The Duflo splitting. Let $\mathfrak{u}$ be a nilpotent Lie algebra and let $\delta \in \mathfrak{u}^*$. Denote by $\pi: U(\mathfrak{u}) \to U(\mathfrak{u})/I_{\mathfrak{u}}(\delta)$ the canonical map. Let $\mathfrak{s}$ be a Lie algebra operating by derivations on $\mathfrak{u}$ and fixing δ.

6.1. PROPOSITION (Duflo's θ-map). *There exists a canonical Lie algebra homomorphism*

$$\theta_\delta: \mathfrak{s} \to U(\mathfrak{u})/I_{\mathfrak{u}}(\delta)$$

having the following properties. Let $\mathfrak{g} := \mathfrak{s} \ltimes \mathfrak{u}$ be the semidirect product. Then

i) *the linear map of $\mathfrak{g}$ into $U(\mathfrak{u})/I_{\mathfrak{u}}(\delta)$ which coincides with θ_δ on $\mathfrak{s}$ and with π on $\mathfrak{u}$ extends to a homomorphism whose kernel will be denoted by I, from $U(\mathfrak{g})$ onto $U(\mathfrak{u})/I_{\mathfrak{u}}(\delta)$.*

ii) *Let $f \in \mathfrak{g}^*$ be the linear form on $\mathfrak{g}$ extending δ which is zero on $\mathfrak{s}$. If $\mathfrak{h}$ is a solvable subalgebra of $\mathfrak{g}$ containing $\mathfrak{u}$, then we have $I \cap U(\mathfrak{h}) = J_{\mathfrak{h}}(f_{|\mathfrak{h}})$.*

For the proof, see ([**Du**, 1] or [**Di**, 5, 10.1]).

6.2. REMARK. The construction of Duflo's θ-map uses in an essential way the Dixmier map for solvable Lie algebras.

6.3. REMARK. From the properties (i) and (ii) in Proposition 6.1, one obtains easily the unicity of θ_δ and, as a consequence, the compatibility with automorphisms of the situation.

If $x \in \mathfrak{u}$, then for the rest of this paragraph we denote by $\overline{x}$ its image in $U(\mathfrak{u})/I_{\mathfrak{u}}(\delta)$.

6.4. REMARK. Let $y \in \mathfrak{s}$ operate on $\mathfrak{u}$ as an inner derivation $\operatorname{ad}(x)$, $x \in \mathfrak{u}$. Then we have $\theta_\delta(y) = \overline{x} - \delta(x)$.

PROOF. Let $\mathfrak{h} = ky \oplus u \subseteq \mathfrak{g}$. Then $y - x$ is a central element of $\mathfrak{h}$. Hence $y - x - f(y-x) \in J_{\mathfrak{h}}(f_{|\mathfrak{h}})$ (see Remark 5.7). This implies $y - x + \delta(x) \in I$ and $\theta_\delta(y) = \overline{x} - \delta(x)$.

6.5. Let now $\mathfrak{u}$ be a nilpotent ideal of a Lie algebra $\mathfrak{g}$, $\delta \in \mathfrak{u}^*$, and suppose that $\mathfrak{g} = \mathfrak{g}(\delta) + \mathfrak{u}$. In this situation, $U(\mathfrak{g})I_{\mathfrak{u}}(\delta)$ is a two-sided (prime) ideal of $U(\mathfrak{g})$.

Let us denote by $\pi: U(\mathfrak{g}) \to U(\mathfrak{g})/U(\mathfrak{g})I_{\mathfrak{u}}(\delta)$ the canonical projection. Let $\mathfrak{s}$ be a subalgebra of $\mathfrak{g}(\delta)$ such that we have still $\mathfrak{g} = \mathfrak{s} + \mathfrak{u}$. Let $M_{\mathfrak{s},\delta}$ be the $\mathfrak{s}$-stable maximal ideal of $U(\mathfrak{s} \cap \mathfrak{u})$ generated by all $x - \delta(x)$ for $x \in \mathfrak{s} \cap \mathfrak{u}$. If $x \in \mathfrak{s}$, we denote by $\overline{\overline{x}}$ its image in $U(\mathfrak{s})/U(\mathfrak{s})M_{\mathfrak{s},\delta}$.

6.6. PROPOSITION (Duflo splitting) (cf. [**Du**, 1] or [**Du**, 5, Ch. IV]). *We have a canonical isomorphism (Duflo splitting)*

$$s_\delta: U(\mathfrak{g})/U(\mathfrak{g})I_{\mathfrak{u}}(\delta) \xrightarrow{\sim} U(\mathfrak{s})/U(\mathfrak{s})M_{\mathfrak{s},\delta} \otimes U(\mathfrak{u})/I_{\mathfrak{u}}(\delta)$$

given by

$$s_\delta(\pi(x)) = 1 \otimes \overline{x} \quad \text{for} \quad x \in \mathfrak{u},$$

$$s_\delta(\pi(x)) = \overline{\overline{x}} \otimes 1 + 1 \otimes \theta_\delta(x) \quad \text{for} \quad x \in \mathfrak{s}.$$

The remark 6.3 guarantees that s_δ is well defined. It is easy to establish an inverse homomorphism so that s_δ is indeed an isomorphism.

6.7. DEFINITION (Duflo's notation) (cf. [**Du**, 5, Ch. IV]). If ξ is a two-sided ideal of $U(\mathfrak{s})$ containing $M_{\mathfrak{s},\delta}$, then we put

$$\xi \circ I_{\mathfrak{u}}(\delta) := \pi^{-1}(s_\delta^{-1}(\xi/(U(\mathfrak{s})M_{\mathfrak{s},\delta}) \otimes U(\mathfrak{u})/I_{\mathfrak{u}}(\delta))).$$

This notation is very suggestive. We mention that there is a slight misuse in the notation since the result depends on δ (and not only on $I_{\mathfrak{u}}(\delta)$). Hence the δ must always be evident from the notation.

6.8. COROLLARY. *The map $\xi \to \xi \circ I_{\mathfrak{u}}(\delta)$ is a bijection between the two-sided ideals of $U(\mathfrak{s})$ containing $M_{\mathfrak{s},\delta}$ and the two-sided ideals of $U(\mathfrak{g})$ containing the maximal ideals $I_{\mathfrak{u}}(\delta)$ of $U(\mathfrak{u})$. The ideal ξ is a rational ideal (hence a primitive ideal) if and only if $\xi \circ I_{\mathfrak{u}}(\delta)$ is a rational ideal (hence a primitive ideal).*

The proof is quite evident using the fact that $U(\mathfrak{u})/I_{\mathfrak{u}}(\delta)$ is a Weyl algebra; in particular, $U(\mathfrak{u})/I_{\mathfrak{u}}(\delta)$ is a simple algebra with center k. It is also quite easy to see that the rationality is preserved. If we want to see that "$\xi \circ I_{\mathfrak{u}}(\delta)$ primitive" implies "ξ primitive", it seems that we need the Dixmier–Moeglin equivalence of rational ideals and primitive ideals.

7. Algebraic Lie algebras. In the following we shall use algebraic Lie algebras and for this purpose we make the terminology precise.

7.1. DEFINITION. An algebraic Lie algebra is a Lie algebra $\mathfrak{g}$ together with two subsets $\mathrm{Un}(\mathfrak{g})$ and $\mathrm{To}(\mathfrak{g})$ of $\mathfrak{g}$ such that there exists a linear algebraic group G over k with the following properties:

i) $\mathfrak{g}$ is the Lie algebra of G;

ii) $x \in \mathrm{Un}(\mathfrak{g})$ (resp. $x \in \mathrm{To}(\mathfrak{g})$) if and only if there is a one-dimensional unipotent subgroup (resp. a one-dimensional torus of G) whose Lie algebra contains x.

From this structure the usual terminology and the usual notations can be recovered. For example, if G is as above, then a subalgebra $\mathfrak{t}$ of $\mathfrak{g}$ is the Lie algebra of a subtorus of G if and only if $\mathfrak{t}$ is generated as a vector space by its intersection with $\mathrm{To}(\mathfrak{g})$ ($\mathfrak{t}$ is then called a torus of $\mathfrak{g}$). A subalgebra $\mathfrak{h}$ of $\mathfrak{g}$ is the Lie algebra of an algebraic subgroup of G ($\mathfrak{h}$ is then called an algebraic subalgebra of $\mathfrak{g}$) if and only if $\mathfrak{h}$ is of the form $\mathfrak{h} = (\mathfrak{s} \oplus \mathfrak{t}) \ltimes \mathfrak{u}$ where $\mathfrak{s}$ is semisimple, $\mathfrak{t}$ is a torus, and $\mathfrak{u}$ is the largest ideal of $\mathfrak{h}$ contained in $\mathrm{Un}(\mathfrak{g})$ (called the unipotent radical of $\mathfrak{h}$ and denoted by ${}^{\mathrm{u}}\mathfrak{h}$).

By a morphism $\varphi: \mathfrak{g}_1 \to \mathfrak{g}_2$ of algebraic Lie algebras we understand a homomorphism of Lie algebras such that $\varphi(\mathrm{Un}(\mathfrak{g}_1)) \subseteq \mathrm{Un}(\mathfrak{g}_2)$ and $\varphi(\mathrm{To}(\mathfrak{g}_1)) \subseteq \mathrm{To}(\mathfrak{g}_2)$. The differential of a homomorphism between linear algebraic groups is such a morphism (for example, ad as the differential of Ad). Let us mention that a bijective morphism φ is an isomorphism, i.e., we have $\varphi(\mathrm{Un}(\mathfrak{g}_1)) = \mathrm{Un}(\mathfrak{g}_2)$ and $\varphi(\mathrm{To}(\mathfrak{g}_1)) = \mathrm{To}(\mathfrak{g}_2)$.

The constructions in the next paragraph will be compatible with such isomorphisms.

If G is a linear algebraic k-group, then we denote by ${}^{\mathrm{u}}G$ its unipotent radical. Hence we have ${}^{\mathrm{u}}\mathfrak{g} = \mathrm{Lie}({}^{\mathrm{u}}G)$ and the notations are compatible. If $\mathfrak{g}$ is an algebraic Lie algebra, then $\mathrm{ad}(\mathfrak{g})$ is algebraic. The adjoint algebraic group Γ is then just the adjoint group, i.e., we have $\Gamma = \mathrm{Ad}(G^0)$ (where G^0 denotes the identity component) for any linear algebraic group G with $\mathfrak{g} = \mathrm{Lie}\, G$ (equality in the sense of algebraic Lie algebras).

7.2. REMARK/DEFINITION. Let $\mathfrak{u}$ be a unipotent subalgebra of an algebraic Lie algebra $\mathfrak{g}$ (i.e., $\mathfrak{u} \subseteq \mathrm{Un}(\mathfrak{g})$). Then there is (up to a canonical isomorphism) a unique unipotent algebraic k-group U such that $\mathfrak{u} = \mathrm{Lie}\, U$. We call U the unipotent group corresponding to $\mathfrak{u}$.

8. Canonical subgroup and canonical subalgebra.

8.1. Let G be a linear algebraic group over k, $\mathfrak{g}$ its Lie algebra and let $f \in \mathfrak{g}^*$. We denote by δ the restriction of f to ${}^{\mathrm{u}}\mathfrak{g}$.

If Y is a subset of a set on which G operates, we denote by $G(Y)$ the stabilizer $\{y \in G | \gamma Y = Y\}$ of Y in G. If G and hence $\mathfrak{g}$ operates linearly on a vector space and if v is an element in this vector space, we put $\mathfrak{g}(v) := \{x \in \mathfrak{g} | xv = 0\}$; hence we have $\mathfrak{g}(v) = \mathrm{Lie}\, G(v)$.

8.2. We now define an operation of reduction: $\mathrm{Red}(G, f)$ is the couple (G_1, f_1) where $G_1 = G(\delta){}^{\mathrm{u}}G$ and where f_1 is the restriction of f to the Lie algebra $\mathfrak{g}_1 = \mathfrak{g}(\delta) + {}^{\mathrm{u}}\mathfrak{g}$ of G_1. Let us mention that G_1 is the stabilizer in G of the (closed) orbit ${}^{\mathrm{u}}G\delta$ in $({}^{\mathrm{u}}\mathfrak{g})^*$.

Let (G_l, f_l) be the result of the iterated reduction $\mathrm{Red}^l(G, f)$. Then we have $G = G_0 \supseteq G_1 \supseteq G_2 \supseteq \cdots$ and f_l is the restriction of f to the Lie algebra g_l of G_l. It is clear that this procedure stops. If $G_l = G_{l+1}$, then we have $\mathrm{Red}^k(G, f) = \mathrm{Red}^l(G, f)$ for all $k \geq l$.

8.3. DEFINITION. We denote the stationary term $G_l = G_{l+1} = \cdots$ by C_f and we denote by $\mathfrak{c}_f$ its Lie algebra. C_f is called the canonical (sub)group of f in G, $\mathfrak{c}_f$ is called the the canonical algebra of f.

8.4. REMARK. We can also define $\operatorname{Red}(\mathfrak{g}, f) := (\mathfrak{g}_1, f_1)$. Let $(\mathfrak{g}_l, f_l) = \operatorname{Red}^n(\mathfrak{g}, f)$. Then we have $\mathfrak{g}_l = \operatorname{Lie} G_l$ and $\mathfrak{c}_f$ is the stationary term of the sequence $\mathfrak{g} = \mathfrak{g}_0 \supseteq \mathfrak{g}_1 \supseteq \cdots$. In particular, $\mathfrak{c}_f$ depends only on the algebraic Lie algebra $\mathfrak{g}$ and on f; it is clear that $\mathfrak{c}_f$ is algebraic.

8.5. PROPOSITION. *Let δ_f be the restriction of f to ${}^{\mathrm{u}}\mathfrak{c}_f$. Then we have:*
i) $\mathfrak{c}_f = \mathfrak{c}_f(\delta_f) + {}^{\mathrm{u}}\mathfrak{c}_f$;
ii) *$\mathfrak{c}_f(\delta_f)$ is the f-orthogonal of ${}^{\mathrm{u}}\mathfrak{c}_f$.*
In particular, $\mathfrak{c}_f$ is f-coisotropic and we have $\mathfrak{g}(f) \subseteq \mathfrak{c}_f(\delta_f)$.

PROOF. The first statement is just the condition for stationarity. With the above notations we have:

$$ {}^{\mathrm{u}}\mathfrak{c}_f = {}^{\mathrm{u}}\mathfrak{g}_l \supseteq {}^{\mathrm{u}}\mathfrak{g}_{l-1} \supseteq \cdots \supseteq {}^{\mathrm{u}}\mathfrak{g} $$

for l large enough. Let $\delta_i := f_{|{}^{\mathrm{u}}\mathfrak{g}_i}$. If $x \in \mathfrak{g}$ and $f([x, {}^{\mathrm{u}}\mathfrak{c}_f]) = 0$, then we get by induction $x \in g_i(\delta_i) \subseteq g_{i+1}$ for any i.

8.6. REMARK. Let $f, f' \in \mathfrak{g}^*$. Suppose that f and f' have the same restriction to ${}^{\mathrm{u}}\mathfrak{c}_f$. Then we have $\mathfrak{c}_f = \mathfrak{c}_{f'}$ (and $C_f = C_{f'}$ with respect to G if $\mathfrak{g} = \operatorname{Lie} G$).

Indeed, if $(\mathfrak{g}'_n, f'_n) = \operatorname{Red}^n(\mathfrak{g}, f)$, then we have $\mathfrak{g}'_n = \mathfrak{g}_n$ for any n; so $\mathfrak{c}_f = \mathfrak{c}_{f'}$ (and similarly $C_f = C_{f'}$).

In order to finish the injectivity proof in §14, we shall need:

8.7. LEMMA. *Let $f \in \mathfrak{g}^*$. Let $\mathfrak{u}$ be the unipotent radical of $\mathfrak{g}$ and U the corresponding unipotent group. Let $\delta := f_{|\mathfrak{u}}$ and $\mathfrak{g}_1 = \mathfrak{g}(\delta) + \mathfrak{g}$. Then we have $f + \mathfrak{g}_1^\perp = U(\delta)f$.*

PROOF. If $\mathfrak{a}$ is a linear subspace of $\mathfrak{g}$, we recall that $\mathfrak{a}^\perp$ denotes its orthogonal in $\mathfrak{g}^*$ and $\mathfrak{a}^{f\perp} := \{x \in \mathfrak{g} | f([x, \mathfrak{a}]) = 0\}$ denotes its f-orthogonal in $\mathfrak{g}$.

i) Let $f_1 := f_{|\mathfrak{g}_1}$. Then we have $f([\mathfrak{u}(\delta), \mathfrak{g}_1]) = 0$; hence $\mathfrak{u}(\delta) f_1 = 0$, $U(\delta) f_1 = f_1$, and $U(\delta) f \subseteq f + \mathfrak{g}_1^\perp$.

ii) We have

$$ \begin{aligned} \mathfrak{u}^{f\perp} &= \mathfrak{g}(\delta) \\ \mathfrak{u}(\delta) &= \mathfrak{u} \cap \mathfrak{u}^{f\perp} \\ \mathfrak{u}(\delta)^{f\perp} &= \mathfrak{u}^{f\perp} + \mathfrak{u} = \mathfrak{g}_1. \end{aligned} $$

This implies $\mathfrak{u}(\delta) f = \mathfrak{g}_1^\perp$ and hence $\dim(U(\delta)f) = \dim \mathfrak{g}_1^\perp$. Since U is unipotent, $U(\delta)f$ is closed in $\mathfrak{g}^*$ and we have $U(\delta)f = f + \mathfrak{g}_1^\perp$.

9. Forms of unipotent type. Let $\mathfrak{g}$ be an algebraic Lie algebra and let Γ be its adjoint group. Any primitive ideal of $U(\mathfrak{g})$ will later on be described (see §14) by its Duflo parameters (up to Γ-conjugation). The first parameter will be a form of unipotent type. These forms, of which we will give an axiomatic description, have been introduced by Duflo (cf. [**Du**, 5, I.10]).

9.1. PROPOSITION. *There exists a unique application Ut $(= Ut_{\mathfrak{g}})$,*

$$ \mathfrak{h} \mapsto Ut(\mathfrak{h}^*), $$

which associates to each algebraic subalgebra $\mathfrak{h}$ *of* $\mathfrak{g}$ *a subset* $Ut(\mathfrak{h}^*)$ *of* $\mathfrak{h}^*$ *with the following properties:*

A0) $0 \in Ut(\mathfrak{h}^*)$;

A1) *If* $f \in \mathfrak{h}^*$ *is such that* $f([\mathfrak{h}, {}^{\mathrm{u}}\mathfrak{h}]) = 0$, *then* $f \in Ut(\mathfrak{h}^*) \Leftrightarrow f(\mathfrak{h}_1) = 0$ *for any reductive subalgebra* $\mathfrak{h}_1$ *of* $\mathfrak{h}$;

A2) *If* $f \in \mathfrak{h}^*$ *and if* $\delta := f_{|{}^{\mathrm{u}}\mathfrak{h}}$, *then* $f \in Ut(\mathfrak{h}^*) \Leftrightarrow f_{|\mathfrak{h}(\delta)} \in Ut(\mathfrak{h}(\delta)^*)$.

9.2. REMARK. If $\mathfrak{g}_1$ is an algebraic subalgebra of $\mathfrak{g}$, then it is clear that the restriction of $Ut_{\mathfrak{g}}$ to the algebraic subalgebras of $\mathfrak{g}_1$ coincides with $Ut_{\mathfrak{g}_1}$. Hence there is no dependence on $\mathfrak{g}$ and the notation Ut (without subscript) is well justified.

Proof of 9.1. We prove the existence and unicity of $Ut_{\mathfrak{g}}$ by induction on the dimension of $\mathfrak{g}$. If $\dim \mathfrak{g} = 0$, then there is nothing to prove.

Hence we may suppose the existence and unicity of $Ut_{\mathfrak{h}}$ for all proper algebraic subalgebras $\mathfrak{h}$ of $\mathfrak{g}$.

If $\mathfrak{h} \subset \mathfrak{g}$ is proper, then we define $Ut_{\mathfrak{g}}(\mathfrak{h}^*) := Ut_{\mathfrak{h}}(\mathfrak{h}^*)$. Then the properties A0), A1), A2) are clear for all such $\mathfrak{h}$.

The case $\mathfrak{h} = \mathfrak{g}$ remains. We define two subsets of $\mathfrak{g}^*$:

$$\Sigma_0(\mathfrak{g}^*) := \{f \in \mathfrak{g}^* | f([\mathfrak{g}, {}^{\mathrm{u}}\mathfrak{g}]) = 0 \text{ and } f(\mathfrak{r}) = 0 \text{ for any reductive subalgebra } \mathfrak{r} \text{ of } \mathfrak{g}\}$$

$$\Sigma_1(\mathfrak{g}^*) := \{f \in \mathfrak{g}^* | f([\mathfrak{g}, {}^{\mathrm{u}}\mathfrak{g}]) \neq 0 \text{ and } f_{|\mathfrak{g}(\delta)} \in Ut_{\mathfrak{g}(\delta)}((\mathfrak{g}(\delta))^*\} \ (\delta := f_{|{}^{\mathrm{u}}\mathfrak{g}}).$$

We put $Ut_{\mathfrak{g}}(\mathfrak{g}) := \Sigma_0(\mathfrak{g}^*) \cup \Sigma_1(\mathfrak{g}^*)$. Then it is clear that $Ut_{\mathfrak{g}}$ is unique and satisfies the properties A0, A1, and A2.

9.3. DEFINITION. We put $\mathfrak{g}^*_{ut} := Ut_{\mathfrak{g}}(\mathfrak{g}^*)$. The elements of $\mathfrak{g}^*_{ut}$ are called forms of *unipotent type* on $\mathfrak{g}$.

9.4. COROLLARY. *If* $\mathfrak{g}$ *is unipotent, then* $\mathfrak{g}^*_{ut} = \mathfrak{g}^*$; *if* $\mathfrak{g}$ *is reductive, then* $\mathfrak{g}^*_{ut} = 0$.

9.5. LEMMA. *Let* $\mathfrak{v}$ *be a unipotent ideal of* $\mathfrak{g}$, $\delta \in \mathfrak{v}^*$, *and suppose that* $\mathfrak{g} = \mathfrak{g}(\delta) + \mathfrak{v}$ *(splitting situation). Let* $f \in \mathfrak{g}^*$ *be a linear form on* $\mathfrak{g}$ *extending* δ. *Then* f *is of unipotent type if and only if its restriction to* $\mathfrak{g}(\delta)$ *is of unipotent type.*

PROOF. Let $\mathfrak{u}$ be the unipotent radical of $\mathfrak{g}$. Let $\mathfrak{w} := \mathfrak{g}(\delta) \cap \mathfrak{u}$, $\delta_0 := f_{|\mathfrak{u}}$ and $\delta' := f_{|\mathfrak{w}}$. Then $\mathfrak{w}$ is the unipotent radical of $\mathfrak{g}(\delta)$ and we have $\mathfrak{g}(\delta_0) = (\mathfrak{g}(\delta))(\delta')$. The property A2 applied to $\mathfrak{g}$ and to $\mathfrak{g}(\delta)$ gives immediately the equivalence of the two properties.

9.6. COROLLARY. *Let* $f \in \mathfrak{g}^*$. *Then* f *is of unipotent type if and only if the restriction of* f *to its canonical algebra* $\mathfrak{c}_f$ *is of unipotent type.*

In this case, we have $\mathfrak{c}_f = \mathfrak{g}(f) + {}^{\mathrm{u}}\mathfrak{c}_f$ and f is zero in any reductive part of $\mathfrak{g}(f)$.

9.7. COROLLARY. *Let $f \in \mathfrak{g}^*$ and let δ_f be the restriction of f to the unipotent radical ${}^{\mathrm{u}}\mathfrak{c}_f$ of the canonical algebra $\mathfrak{c}_f$. Then the following are equivalent:*

i) *f is of unipotent type;*

ii) *the restriction of f to $\mathfrak{c}_f$ is of unipotent type;*

iii) *the restriction of f to $\mathfrak{c}_f(\delta_f)$ is of unipotent type;*

iv) *f is zero on any reductive part of $\mathfrak{c}_f(\delta_f)$;*

v) *$\mathfrak{c}_f = \mathfrak{g}(f) + {}^{\mathrm{u}}\mathfrak{c}_f$ and f is zero on any reductive part of $\mathfrak{g}(f)$.*

PROOF. Let $(\mathfrak{g}_i, f_i) = \mathrm{Red}^i(\mathfrak{g}, f)$. Let $\delta_i := f_{|{}^{\mathrm{u}}\mathfrak{g}_i}$, $\mathfrak{g}'_{i+1} = \mathfrak{g}_i(\delta_i)$, $f'_i := f_{i|\mathfrak{g}'_i}$. Hence $\mathfrak{g}_{i+1} = \mathfrak{g}'_{i+1} + u_i$ where $u_i := {}^{\mathrm{u}}\mathfrak{g}_i$. From 10.4 we get that f_i $(= f_{|\mathfrak{g}_i})$ is of unipotent type if and only if f is of unipotent type.

9.8. REMARK. Let $f \in \mathfrak{g}^*_{ut}, \delta_f, \mathfrak{c}_f$ be as in Cor. 9.7. Since $\mathfrak{c}_f$ is f-coisotropic and $\mathfrak{c}_f = \mathfrak{g}(f) + {}^{\mathrm{u}}\mathfrak{c}_f$, it is easy to see that $f + \mathfrak{c}_f^{\perp} \subseteq Vf$ if V is the unipotent group corresponding to ${}^{\mathrm{u}}\mathfrak{c}_f$. Hence, if f' is another form of unipotent type extending δ_f, then f and f' are $\Gamma_{\mathfrak{g}}$-conjugate.

The property 9.7(v) can be used to characterize forms of unipotent type without using any recurrence procedure.

9.9. DEFINITION ([**Du**, 5, I.7]). Let $f \in \mathfrak{g}^*$. A subalgebra $\mathfrak{b} \subset \mathfrak{g}$ is called of *strongly unipotent type* if

i) $\mathfrak{b}$ is algebraic,

ii) $\mathfrak{b}$ is f-coisotropic,

iii) $\mathfrak{b} = \mathfrak{g}(f) + {}^{\mathrm{u}}\mathfrak{b}$.

We give the following proposition without proof (cf. [**Du**, 5, Ch. I]).

9.10. PROPOSITION. *Let $f \in \mathfrak{g}^*$. Then f is of unipotent type if and only if it satisfies the following conditions:*

U1) *f is zero on any reductive part of $\mathfrak{g}(f)$;*

U2) *there exists a subalgebra of strongly unipotent type for f.*

From 10.4 we get the one direction, that an element of unipotent type satisfies U1 and U2. For the converse, we refer to Duflo [**Du**, Ch. 1]. In particular, if $\mathfrak{g}$ is reductive, then U1 and U2 imply $f = 0$ [**Du**, 5, I.13]. Duflo uses U1 and U2 as defining properties. There is also the notion of a subalgebra of unipotent type (with respect to f). If $f \in \mathfrak{g}^*_{ut}$, then the two notions coincide.

9.11. COROLLARY. *Let $f \in \mathfrak{g}^*_{ut}$. Let $\mathfrak{h}$ be an algebraic subalgebra of $\mathfrak{g}$ containing the canonical subalgebra $\mathfrak{c}_f$ (or containing any subalgebra of (strongly) unipotent type for f). Then $f_{|\mathfrak{h}}$ is a form of unipotent type on $\mathfrak{h}$.*

This corollary results from Duflo's characterization: it seems that it is not an easy consequence of the axioms A0–A2.

10. Good position. Further notations. Let $\mathfrak{g}$ be an algebraic Lie algebra.

10.1. DEFINITION (cf. [**Du**, 5, IV.1]). Let $f \in \mathfrak{g}^*$. Then we define $Z(f)$ to be the set of (two-sided) ideals of $U(\mathfrak{g}(f))$ containing all $x - f(x)$ for $x \in {}^{\mathrm{u}}(\mathfrak{g}(f))$.

10.2. REMARK. Let $\mathfrak{r}$ be a Levi factor of $\mathfrak{g}(f)$ and $\bar{\mathfrak{r}} := \mathfrak{g}(f)/^{\mathrm{u}}(\mathfrak{g}(f))$ $(\cong \mathfrak{r})$. Then it is clear that $Z(f)$ can be canonically identified (by taking intersections) with the two-sided ideals of $U(\mathfrak{r})$ $(\cong U(\bar{\mathfrak{r}}))$.

10.3. DEFINITION. Let $\mathfrak{g}_1 \subseteq \mathfrak{g}$ be an algebraic subalgebra of $\mathfrak{g}$. Let $f \in \mathfrak{g}^*$ and $f_1 \in \mathfrak{g}_1^*$. We say that (f, f_1) is in good position if

i) $f_1 = f_{|\mathfrak{g}_1}$;

ii) $\mathfrak{g}(f) \subseteq \mathfrak{g}_1(f_1)$;

iii) $\mathfrak{g}_1(f_1) = \mathfrak{g}(f) + {}^{\mathrm{u}}(\mathfrak{g}_1(f_1))$.

If this is the case and if $\xi \in Z(f)$, then we denote by $\xi_{\mathrm{ex}}(\in Z(f_1))$ the ideal generated by ξ and by all $x - f(x)$ for $x \in {}^{\mathrm{u}}(\mathfrak{g}_1(f_1))$. (In this notation, f_1 is supposed to be clear from the context.)

10.4. REMARK. Let $\mathfrak{g}_2 \subset \mathfrak{g}_1 \subset \mathfrak{g}$ be algebraic subalgebras, $f \in \mathfrak{g}^*$, $f_1 := f_{|\mathfrak{g}_1}$, $f_2 := f_{|\mathfrak{g}_2}$. If (f, f_1) and (f, f_2) are in good position, then (f, f_2) is in good position and we have canonical (and inverse) bijections:

$$Z(f_2) \to Z(f_1) \to Z(f) \text{ and } Z(f) \overset{\mathrm{ex}}{\to} Z(f_1) \overset{\mathrm{ex}}{\to} Z(f_2)$$

(where on the left the maps are just "taking intersections").

10.5. REMARK. Let $\mathfrak{v}$ be a unipotent ideal of $\mathfrak{g}$, $\delta := f_{|\mathfrak{v}}$, $\mathfrak{g}_1' := \mathfrak{g}(\delta)$, $f_1' := f_{|\mathfrak{g}_1'}$, $\mathfrak{g}_1 := \mathfrak{g}(\delta) + \mathfrak{v}$, $f_1 := f_{|\mathfrak{g}_1}$. Then we have $\mathfrak{g}_1(f_1) = \mathfrak{g}_1'(f_1')$ and (f, f_1) and (f, f_1') are in good position.

PROOF. i) Since $\mathfrak{g}_1(f_1) \subseteq \mathfrak{g}(\delta) = \mathfrak{g}_1'$ and $f([\mathfrak{g}_1'(f_1'), \mathfrak{g}_1' + \mathfrak{v}]) = 0$, we have $\mathfrak{g}_1(f_1) = \mathfrak{g}_1'(f_1')$.

ii) We have $\mathfrak{g}_1'(f_1') = \mathfrak{g}(f) + \mathfrak{v}(\delta)$ (cf. [**Du**, 5, I.16]). Indeed, we have $\mathfrak{g}_1' = \mathfrak{v}^{f\perp}$. Hence

$$(\mathfrak{g}_1')^{f\perp} = \mathfrak{v} + \mathfrak{g}(f),$$
$$\mathfrak{g}_1'(f_1') = \mathfrak{g}_1' \cap (\mathfrak{v} + \mathfrak{g}(f)) = \mathfrak{v}(\delta) + \mathfrak{g}(f).$$

10.6. NOTATION. Let $f \in \mathfrak{g}^*$, $\mathfrak{u} := {}^{\mathrm{u}}\mathfrak{g}$, $\delta := f_{|\mathfrak{u}}$. Let $\mathfrak{g}_1 := \mathfrak{g}(\delta) + \mathfrak{u}$, $f_1 := f_{|\mathfrak{g}_1}$. We recall the definition (8.3) $\mathrm{Red}(\mathfrak{g}, f) := (\mathfrak{g}_1, f_1)$ and we extend this definition by

$$\mathrm{Red}(\mathfrak{g}, f, \xi) := (\mathfrak{g}_1, f_1, \xi_1)$$

where $\xi \in Z(f)$ and $\xi_1 := \xi_{\mathrm{ex}} \subseteq U(\mathfrak{g}_1(f_1))$. ($\xi_{\mathrm{ex}}$ = the two-sided ideal of $U(\mathfrak{g}_1(f_1))$ generated by ξ and the $x - f(x)$ for $x \in \mathfrak{u}(\delta)$).

11. The Duflo construction.

11.1. Let $\mathfrak{g}$ be an algebraic Lie algebra (see §7). If $f \in \mathfrak{g}_{ut}^*$, i.e., if f is a form of unipotent type, then we have (Cor. 9.7) for the canonical algebra of f the following situation:

$$\mathfrak{c}_f = \mathfrak{g}(f) + {}^{\mathrm{u}}(\mathfrak{c}_f).$$

This implies, in particular, that

$$\mathfrak{g}(f) \cap {}^{\mathrm{u}}(\mathfrak{c}_f) = {}^{\mathrm{u}}(\mathfrak{g}(f)).$$

If $\xi \in Z(f)$ (see 10.1), i.e., if ξ is a (two-sided) ideal of $U(\mathfrak{g}(f))$ containing all $x - f(x)$ for $x \in {}^{\mathrm{u}}(\mathfrak{g}(f))$, then we can define the ideal of $\mathfrak{c}_f$,

$$\xi \circ I_{\mathfrak{v}_f}(\delta_f),$$

where $\mathfrak{v}_f := {}^{\mathrm{u}}(\mathfrak{c}_f)$ and $\delta_f := f_{|\mathfrak{v}_f}$.

11.2. DEFINITION. We define for $f \in \mathfrak{g}^*_{ut}$ and $\xi \in Z(f)$ (cf. [**Du**, 5, IV]),

$$I_{\mathfrak{g}}(f, \xi) := \mathrm{Ind}^{\sim}(\xi \circ I_{\mathfrak{v}_f}(\delta_f); \mathfrak{c}_f \uparrow \mathfrak{g}).$$

There is a remarkable independence statement saying that we can replace the canonical algebra $\mathfrak{c}_f$ by any other subalgebra of (strongly) unipotent type (see §9) with respect to f.

11.3. THEOREM ([**Du**, 5, IV.8]). *Let $f \in \mathfrak{g}^*_{ut}$, $\xi \in Z(f)$ and let $\mathfrak{b} = \mathfrak{g}(f) + {}^{\mathrm{u}}\mathfrak{b}$ be any subalgebra of (strongly) unipotent type with respect to f. Let $\mathfrak{v} := {}^{\mathrm{u}}\mathfrak{b}$. Then we have $I_{\mathfrak{g}}(f, \xi) = \mathrm{Ind}^{\sim}(\xi \circ I_{\mathfrak{v}}(f_{|\mathfrak{v}}); \mathfrak{b} \uparrow \mathfrak{g})$.*

12. Some properties of the Duflo map.

12.1. Let $\mathfrak{g}$ be an algebraic Lie algebra. If a is an automorphism of the algebraic Lie algebra $\mathfrak{g}$ then it is clear that

$$aI_{\mathfrak{g}}(f, \xi) = I_{\mathfrak{g}}(af, a\xi),$$

where $f \in \mathfrak{g}^*_{ut}$, $\xi \in Z(f)$ and $af := f \circ a^{-1}$ (since all constructions are compatible with such an automorphism).

Now let $\mathfrak{v}$ be a unipotent ideal of $\mathfrak{g}$, $f \in \mathfrak{g}^*_{ut}$, $\delta := f_{|\mathfrak{v}}$ and $\xi \in Z(f)$. Let $\mathfrak{g}_1 := \mathfrak{g}(\delta) + \mathfrak{v}$, $f_1 := f_{|\mathfrak{g}_1}$, $\mathfrak{g}'_i := \mathfrak{g}(\delta)$, $f'_1 := f_{|\mathfrak{g}'_1}$. Then we know (§10) that $\mathfrak{g}_1(f_1) = \mathfrak{g}'_1(f'_1)$ and that (f, f_1) is in good position. Let $\xi_{\mathrm{ex}} \subseteq U(\mathfrak{g}'_1(f'_1)) = U(\mathfrak{g}_1(f_1))$ be the ideal of $U(\mathfrak{g}_1(f_1))$ generated by ξ and the $x - f(x)$ for $x \in \mathfrak{g}_1(f_1) \cap \mathfrak{v} = {}^{\mathrm{u}}(\mathfrak{g}_1(f_1))$.

With these notations we have

12.2. THEOREM (cf. [**Du**, 5, IV.9]).

$$I_{\mathfrak{g}}(f, \xi) = \mathrm{Ind}^{\sim}(I_{\mathfrak{g}'_1}(f'_1, \xi_{\mathrm{ex}}) \circ I_v(\delta); \mathfrak{g}_1 \uparrow \mathfrak{g}).$$

For the proof, we refer to [**Du**, 5, Ch. IV].

12.3. COROLLARY. *Let Γ be the adjoint algebraic group of $\mathfrak{g}$. If $\mathfrak{v}$ is a unipotent ideal of $\mathfrak{g}$, if $f \in \mathfrak{g}^*_{ut}$ and $\xi \in Z(f)$, then we have*

$$I_{\mathfrak{g}}(f, \xi) \cap U(\mathfrak{v}) = \bigcap_{\gamma \in \Gamma} \gamma I_{\mathfrak{v}}(f_{|\mathfrak{v}}).$$

This is immediate from the induction formula of Thm. 12.2.

12.4. COROLLARY. *In the situation of Thm. 12.2, we have*

i) $I_{\mathfrak{g}_1}(f_1, \xi_{\mathrm{ex}}) = I_{\mathfrak{g}'_1}(f'_1, \xi_{\mathrm{ex}}) \circ I_{\mathfrak{v}}(\delta)$;

ii) $I_{\mathfrak{g}}(f, \xi) = \mathrm{Ind}^{\sim}(I_{\mathfrak{g}_1}(f_1, \xi_{\mathrm{ex}}); \mathfrak{g}_1 \uparrow \mathfrak{g})$.

Indeed, point (i) follows from Thm. 12.2 applied to $(\mathfrak{g}_1, f_1, \xi_{\mathrm{ex}}, \mathfrak{v})$ instead of $(\mathfrak{g}, f, \xi, \mathfrak{v})$. Point (ii) is then an immediate consequence.

For further information concerning induction formulas like (ii) and some more detailed information about such situations, we refer to [**M-R**, 3, Ch. III].

13. The primitivity of the Duflo result.

13.1. PROPOSITION. *Let $\mathfrak{g}$ be an algebraic Lie algebra. Let $f \in \mathfrak{g}^*_{ut}$ and $\xi \in Z(f)$. If ξ is primitive, then $I_{\mathfrak{g}}(f,\xi)$ is primitive (cf.* [**Du**, 5, IV]).

PROOF. Let $u := {}^{\mathrm{u}}\mathfrak{g}$ and $\delta := f_{|u}$. Let $\mathfrak{g}_1' := \mathfrak{g}(\delta)$, $f_1' := f_{|\mathfrak{g}_1'}$. Let $(\mathfrak{g}_1, f_1, \xi_1) := \mathrm{Red}(\mathfrak{g}, f, \xi)$ (§10). Hence $\mathfrak{g}_1 = \mathfrak{g}(\delta) + \mathfrak{u} = \mathfrak{g}_1(\delta) + \mathfrak{u}$, $f_1 = f_{|\mathfrak{g}_1}$, $\xi_1 = \xi_{\mathrm{ex}}$. Let

$$I_1 := I_{\mathfrak{g}_1}(f_1, \xi_1),$$
$$I_1' := I_{\mathfrak{g}_1'}(f_1', \xi_1).$$

Then we know from §12 that

$$I_1 = I_1' \circ I_{\mathfrak{u}}(\delta),$$
$$I = \mathrm{Ind}^{\sim}(I_1, \mathfrak{g}_1 \uparrow \mathfrak{g}).$$

We proceed by induction on the dimension of $\mathfrak{g}$. If $\mathfrak{g} = \mathfrak{g}_1'$, i.e., $\mathfrak{g} = \mathfrak{g}(\delta)$ then f is the unique element of unipotent type extending δ and we have $\mathfrak{g} = \mathfrak{g}(f) = \mathfrak{c}_f$ and $I = \xi$; hence there is nothing to prove. If $\mathfrak{g}(\delta) \neq \mathfrak{g}_1$, then I_1 is primitive by the hypothesis of induction. Let (ρ, W) be an irreducible representation of $\mathfrak{g}_1$ with kernel I_1' in $U(\mathfrak{g}_1)$ and let (σ, V) be an irreducible representation of $\mathfrak{u}$ with kernel $I_{\mathfrak{u}}(\delta)$. Then (cf. [**Du**, 1, Thm. 3.1]) we have an irreducible representation $\tilde{\rho}$ of $\mathfrak{g}$ in the space $W \otimes V$ defined by

$$\tilde{\rho}(x) = \sigma(x) \quad \text{for} \quad x \in \mathfrak{u},$$
$$\tilde{\rho}(y) = (\rho(y) \otimes 1) + (1 \otimes \sigma(\theta_\delta(y)) \quad \text{for} \quad y \in \mathfrak{g}(\delta),$$

where θ_δ is the Duflo θ-map. We have already seen above that for $y \in \mathfrak{u} \cap \mathfrak{g}(\delta)$ the two defining lines give the same result. It is clear that $\tilde{\rho}/\mathfrak{u}$ is a multiple of σ.

In this situation, we can apply Blattner's irreducibility theorem (see [**Di**, 1, Thm. 5.36] or [**B**]). Let (cf. [**Di**, 1, Thm. 5.3.1]) $\mathrm{st}(\sigma, \mathfrak{g})$ denote the subalgebra of all $y \in \mathfrak{g}$ for which there is an $s \in \mathrm{End}_k(V)$ such that $\sigma([y,x]) = [s, \sigma(x)]$. Since $U(\mathfrak{u})/I_{\mathfrak{u}}(\delta)$ is a Weyl algebra (hence all of its derivations are inner), we have $\mathrm{st}(\sigma, \mathfrak{g}) = \mathrm{st}_{\mathfrak{g}}(I_{\mathfrak{u}}(\delta))$ $(= \mathfrak{g}_1)$. Then Blattner's theorem gives the irreducibility of $\mathrm{ind}^{\sim}(\tilde{\rho}; \mathfrak{g}_1 \uparrow \mathfrak{g})$ and hence the primitivity of $I = \mathrm{Ind}^{\sim}(I_1; \mathfrak{g}_1 \uparrow \mathfrak{g})$ using the fact that I_1 is the kernel of $\tilde{\rho}$ in $U(\mathfrak{g})$.

14. The bijectivity of the factored Duflo map. Let $\mathfrak{g}$ be an algebraic Lie algebra and let Γ be its adjoint group. We use the following notation (cf. [**Du**, 5, IV.1]).

14.1. NOTATION. We denote by $\mathcal{F}(\mathfrak{g})$ the set of all couples (f, ξ) where $f \in \mathfrak{g}^*_{ut}$ and where ξ is a primitive ideal in $Z(f)$, i.e., a primitive ideal of $U(\mathfrak{g}(f))$ containing all $x - f(x)$ for $x \in {}^{\mathrm{u}}(\mathfrak{g}(f))$.

In this paragraph we give the following classification theorem (cf. [**Du**, 5, IV.7] for the surjectivity and [**M-R**, 2] for the injectivity).

14.2. THEOREM. i) *Any primitive ideal I of $U(g)$ is of the form $I_{\mathfrak{g}}(f,\xi)$ for some couple $(f,\xi) \in \mathcal{F}(\mathfrak{g})$.*

ii) *If (f,ξ) and (f',ξ') are in $J(\mathfrak{g})$ and if $I_{\mathfrak{g}}(f,\xi) = I_{\mathfrak{g}}(f',\xi')$, then there exists a $\gamma \in \Gamma$ such that $f' = \gamma f$ and $\xi' = \gamma \xi$.*

14.3. DEFINITION. If $(f,\xi) \in J(\mathfrak{g})$ is such that $I = I_{\mathfrak{g}}(f,\xi)$, then f and ξ are called *Duflo parameters* of I (by keeping in mind that the couple (f,ξ) is only defined up to Γ-conjugation).

Let $\mathfrak{u}$ denote the unipotent radical of $\mathfrak{g}$ and let I be a primitive ideal of $U(\mathfrak{g})$. Let $\omega \in \operatorname{Prim} U(\mathfrak{u})$ be the support of I with respect to $\mathfrak{u}$ (see §2), i.e., the unique Γ-orbit of primitive ideals Q of $U(\mathfrak{u})$ having the property that

$$I \cap U(\mathfrak{u}) = \bigcap_{\gamma \in \Gamma} \gamma Q.$$

We fix such a $Q \in \omega$. Let $\delta \in \mathfrak{u}^*$ be such that $Q = I_{\mathfrak{u}}(\delta)$. Let G be a connected linear k-group such that $\mathfrak{g} = \operatorname{Lie} G$ (as algebraic Lie algebra); hence $\Gamma = \operatorname{Ad}(G)$. Let U be the unipotent radical of G. Let $G_1 = G(\delta)U$. Then G_1 is the stabilizer of $U\delta$ in G and hence also the stabilizer of Q. Let $\mathfrak{g}_1 = \operatorname{Lie} G_1$, i.e., $\mathfrak{g}_1 = \mathfrak{g}(\delta) + \mathfrak{u}$. In this situation we have

14.4. PROPOSITION (see [**M-R**, 1, Thm. 4.7]). *There is a primitive ideal J of $U(\mathfrak{g}_1)$ such that $J \cap U(\mathfrak{u}) = Q$ and such that* $\operatorname{Ind}^{\sim}(J; \mathfrak{g}_1 \uparrow \mathfrak{g}) = I$.

(This proposition and also the next one hold even if $\mathfrak{u}$ is any ideal of $\mathfrak{g}$ by taking for G_1 the stabilizer of Q in G.) Moreover, we have the following unicity result.

14.5. PROPOSITION (see [**M-R**, 2, Introduction, and 2.4]). *If I' is another primitive ideal of $U(\mathfrak{g}_1)$ such that $I' \cap U(\mathfrak{u}) = Q$ and such that* $\operatorname{Ind}^{\sim}(I'; \mathfrak{g}_1 \uparrow \mathfrak{g}) = I$, *then there is a $\gamma \in G_1$ such that $I' = \gamma I$.*

We now continue the proof of the theorem. We suppose that the theorem is true for smaller dimensions than $\dim \mathfrak{g}$. We prove (i) and (ii) together. For proving (ii) we put $I = I_{\mathfrak{g}}(f,\xi) = I_{\mathfrak{g}}(f',\xi')$. If $\mathfrak{g} = \mathfrak{g}_1$, then we are in the Duflo splitting situation, i.e., $\mathfrak{g} = \mathfrak{g}(\delta) + \mathfrak{u}$ (cf. §6). Hence the ideal I is of the form $I = I' \circ I_{\mathfrak{u}}(\delta)$, using the Duflo notation (§6) where I' is an ideal of $U(\mathfrak{g}(\delta))$ containing $x - \delta(x)$ for $x \in \mathfrak{g}(\delta) \cap \mathfrak{u}$. Since I is a primitive ideal, I is rational. This implies that I' is rational and by the Dixmier–Moeglin result, we have that I' is primitive. (It seems difficult to see directly that I' is primitive.) If $\mathfrak{g} = \mathfrak{g}(\delta)$, then there is nothing to prove since $I = I_{\mathfrak{g}}(f_0, \xi)$ where f_0 is the unique element of $\mathfrak{g}^*_{ut}$ extending δ, $\mathfrak{g}(f_0) = \mathfrak{g}$ and $\xi = I$. If $\mathfrak{g} = \mathfrak{g}_1$ but $\mathfrak{g} \neq \mathfrak{g}(\delta)$, we know that $I' = I_{\mathfrak{g}(\delta)}(f',\xi')$ with $(f',\xi') \in \mathcal{F}(\mathfrak{g}(\delta))$ and we know that (f',ξ') is unique up to $G(\delta)^0$-conjugation. By taking intersection with $U(\mathfrak{g}(\delta) \cap \mathfrak{u})$ we see that f' and δ coincide on $\mathfrak{g}(\delta) \cap \mathfrak{u}$. Let $f^0 \in \mathfrak{g}^*$ be the linear form extending f' and δ. Then we know (by A2) that $f^0 \in \mathfrak{g}^*_{ut}$ and by §10 that (f^0, f') is in good position. Let $\xi^0 = \xi' \cap U(\mathfrak{g}(f^0))$. Then we have, by the properties of the Duflo map, that

$$I_{\mathfrak{g}}(f^0, \xi^0) = I_{\mathfrak{g}(\delta)}(f',\xi') \circ I_{\mathfrak{u}}(\delta).$$

This proves that $I = I_{\mathfrak{g}}(f^0, \xi^0)$, i.e., the surjectivity. Since (f', ξ') is unique up to $G(\delta)^0$-conjugation, (f^0, ξ^0) is also unique up to $G(\delta)^0$-conjugation once $f^0_{|\mathfrak{u}} = \delta$ is fixed. Hence (f, ξ) is unique up to G-conjugation.

If $\mathfrak{g} \neq \mathfrak{g}_1$, then we know by the induction hypothesis that there is $(f_1, \xi_1) \in \mathcal{F}(\mathfrak{g}_1)$ with $I_1 = I_{\mathfrak{g}_1}(f_1, \xi_1)$ and that (f_1, ξ_1) is unique up to G_1^0-conjugation. We know that $I_1 \cap U(\mathfrak{u}) = I_{\mathfrak{u}}(\delta) = \bigcap_{\gamma \in G_1^0} I_{\mathfrak{u}}(\gamma f_{1|\mathfrak{u}})$. Hence $I_{\mathfrak{u}}(\delta) = I_{\mathfrak{u}}(\gamma f_{1|\mathfrak{u}})$ for all $\gamma \in G_1^0$ and we may assume that $\gamma f_{1|\mathfrak{u}} = \delta$.

Let $f \in \mathfrak{g}^*$ be any extension of f_1. Since $f_{1|\mathfrak{g}(\delta)}$ is of unipotent type by Lemma 9.5, f is of unipotent type by A2. By §10 we know that (f, f_1) is in good position. Let $\xi := \xi_1 \cap U(\mathfrak{g}(f))$. Then by the properties of the Duflo map (§12) we have $I_{\mathfrak{g}}(f, \xi) = \mathrm{Ind}^{\sim}(I_1; \mathfrak{g}_1 \uparrow \mathfrak{g})$; hence $I = I_{\mathfrak{g}}(f, \xi)$. This proves the surjectivity.

If $I = I_{\mathfrak{g}}(f, \xi) = I_{\mathfrak{g}}(\tilde{f}, \tilde{\xi})$ with $(\tilde{f}, \tilde{\xi}) \in \mathcal{F}(\mathfrak{g})$, then we have (12.3): $I \cap U(\mathfrak{u}) = \bigcap_{\gamma \in \Gamma} \gamma I_{\mathfrak{u}}(f_{|\mathfrak{u}}) = \bigcap_{\gamma \in \Gamma} \gamma I_{\mathfrak{u}}(\tilde{f}_{|\mathfrak{u}})$ and by a suitable conjugation (cf. §2) we can assume that $f_{|\mathfrak{u}} = \tilde{f}_{|\mathfrak{u}} = \delta$.

Let $f_1 := f_{|\mathfrak{g}_1}$ and $\tilde{f}_1 := \tilde{f}_{|\mathfrak{g}_1}$, $\xi_1 := \xi_{\mathrm{ex}}$ (from $Z(f) \overset{\mathrm{ex}}{\tilde{\to}} Z(f_1)$) and $\tilde{\xi}_1 := \tilde{\xi}_{\mathrm{ex}}$ (from $Z(\tilde{f}) \overset{\mathrm{ex}}{\tilde{\to}} Z(\tilde{f}_1)$). Then we have

$$I_{\mathfrak{g}_1}(f_1, \xi_1) \cap U(\mathfrak{u}) = I_{\mathfrak{g}_1}(\tilde{f}_1, \tilde{\xi}_1) \cap U(\mathfrak{u}) = I_{\mathfrak{u}}(\delta)$$

and

$$\mathrm{Ind}^{\sim}(I_{\mathfrak{g}_1}(f_1, \xi_1); \mathfrak{g}_1 \uparrow \mathfrak{g}) = \mathrm{Ind}^{\sim}(I_{\mathfrak{g}_1}(\tilde{f}_1, \tilde{\xi}_1), \mathfrak{g}_1 \uparrow \mathfrak{g}) = I.$$

By 14.5 there exists $\gamma \in G_1 = G(\delta)U$ such that $I_{\mathfrak{g}_1}(f_1, \xi_1) = \gamma I_{\mathfrak{g}_1}(\tilde{f}_1, \tilde{\xi}_1)$. We can suppose that $\gamma \in G(\delta)$. As we have $f_{1|\mathfrak{u}} = \tilde{f}_{1|\mathfrak{u}} = \delta$ and as we know that $I_{\mathfrak{g}_1}$ is injective up to G_1^0-conjugation, we can assume that $f_1 = \tilde{f}_1$ and $\xi_1 = \tilde{\xi}_1$. Now by Lemma 8.7, we have $U(\delta) f_1 = f_1 + \mathfrak{g}_1^{\perp}$; hence f and $\tilde{f}$ are G-conjugate (and hence Γ-conjugate). Moreover, we have automatically $\xi = \tilde{\xi}$ after the above conjugation operations. So the proof of the classification theorem is established.

14.6. QUESTION. If $(f, \xi) \in \mathcal{F}(\mathfrak{g})$, do then ξ and $I_{\mathfrak{g}}(f, \xi)$ have the same Goldie rank?

For some partial results to this question, see [**M-R**, 4].

15. The Dixmier–Duflo map in the general case and a decomposition of the primitive spectrum.

15.1. Let $\mathfrak{g}$ be any (finite-dimensional) Lie algebra over k and let Γ be its adjoint algebraic group. We denote by $\mathfrak{g}^*_{\mathrm{sp}}$ the subset of elements of $\mathfrak{g}^*$ having a solvable polarization. As shown by Dixmier (cf. [**Di**, 4]) and Duflo (cf. [**Du**, 3]), $\mathfrak{g}^*_{\mathrm{sp}}$ contains the set of regular elements, i.e., the set of those $f \in \mathfrak{g}^*$ for which $\mathfrak{g}(f)$ is of minimal dimension.

If $f \in \mathfrak{g}^*$ and if $\mathfrak{h}$ is a subalgebra of $\mathfrak{g}$ such that $f([\mathfrak{h}, \mathfrak{h}]) = 0$, we keep the notation $J(f, \mathfrak{h})$ from 5.2, i.e., $J(f, \mathfrak{h})$ is the kernel in $U(\mathfrak{g})$ of the representation $\mathrm{ind}^{\sim}(f_{|\mathfrak{h}}; \mathfrak{h} \uparrow \mathfrak{g})$. Such an ideal is completely prime, since it is induced from a completely prime ideal (cf. [**C**,1]).

Then as shown by Duflo (see [**Du**, 3] or [**Di**, 5, Thm. 10.3.3]) we have:

15.2. THEOREM. *Let* $f \in \mathfrak{g}^*_{\mathrm{sp}}$.

i) *There exists a solvable polarization* $\mathfrak{h}$ *of* f *such that* $\mathrm{ind}^\sim(f_{|\mathfrak{h}}; \mathfrak{h} \uparrow \mathfrak{g})$ *is irreducible.*

ii) *If* $\mathfrak{h}_1$ *and* $\mathfrak{h}_2$ *are two solvable polarizations of* f, *then the kernels* $J(f, \mathfrak{h}_1)$ *and* $J(f, \mathfrak{h}_2)$ *of* $\mathrm{ind}^\sim(f_{|\mathfrak{h}_1}; \mathfrak{h}_1 \uparrow \mathfrak{g})$ *and of* $\mathrm{ind}^\sim(f_{|\mathfrak{h}_2}; \mathfrak{h}_2 \uparrow \mathfrak{g})$ *coincide.*

We would like to mention that the proof uses the Duflo splitting (§6) in an essential way.

15.3. DEFINITION. If $f \in \mathfrak{g}^*_{\mathrm{sp}}$, then we denote by $J(f)$ (or $J_{\mathfrak{g}}(f)$) the primitive completely prime ideal $J(f, \mathfrak{h})$ where $\mathfrak{h}$ is any solvable polarization of f. The map $J: \mathfrak{g}^*_{\mathrm{sp}} \to \mathrm{Prim}\, U(\mathfrak{g})$ is called the Dixmier–Duflo map. The ideals $J_{\mathfrak{g}}(f)$ are called the Duflo ideals of $U(\mathfrak{g})$.

15.4. REMARK. From the construction of $J(f)$ it is quite easy to see that $J_{\mathfrak{g}}$ commutes with automorphisms of $\mathfrak{g}$ and that $J_{\mathfrak{g}}(f + \lambda) = \tau_{-\lambda} J_{\mathfrak{g}}(f)$ for $\lambda \in \mathfrak{g}^*$ with $\lambda([\mathfrak{g}, \mathfrak{g}]) = 0$. In particular, we have $J_{\mathfrak{g}}(\gamma f) = J_{\mathfrak{g}}(f)$ for $\gamma \in \Gamma$.

We mention two important properties, as established by Moeglin (see [**M**]; also for the proof).

15.5. PROPOSITION.

i) *If* $f_1, f_2 \in \mathfrak{g}^*_{\mathrm{sp}}$, *then* $J(f_1) = J(f_2)$ *if and only if there is a* $\gamma \in \Gamma$ *such that* $f_2 = \gamma f_1$.

ii) *Any primitive ideal of* $U(\mathfrak{g})$ *contains a primitive ideal* $J(f)$ *of Duflo for some* $f \in \mathfrak{g}^*_{\mathrm{sp}}$.

If $\mathfrak{g}$ is semisimple, then $\mathfrak{g}^*_{\mathrm{sp}}$ coincides with the set of regular elements (cf. [**Du**, 3, 1.1]) and the solvable polarizations are Borel subalgebras. In this case, the ideals $J(f)$ are just the minimal primitive ideals of $U(\mathfrak{g})$ and they are generated by their intersection with the center of $U(\mathfrak{g})$ (cf. [**Di**, 5, 8.4.3]). For $\mathfrak{g}$ general, there may be inclusion relations between the $J(f)$'s and we don't know whether we can identify the Duflo ideals only from the algebraic structure of $U(\mathfrak{g})$.

However, if we use in addition the natural operation of Γ on $U(\mathfrak{g})$, then we have for algebraic $\mathfrak{g}$ the following characterization of the Duflo ideals (see [**M-R**, 3, III.20]).

15.6. PROPOSITION. *Let* $\mathfrak{g}$ *be the Lie algebra of a connected linear algebraic* k*-group* G. *A primitive ideal* I *of* $U(\mathfrak{g})$ *is an ideal of Duflo (i.e., of the form* $J_{\mathfrak{g}}(f)$*) if and only if there exists a closed subgroup* B *of* G *whose Lie algebra is solvable and a* G*-equivariant homomorphism from the field of rational functions on* G/B *into the total ring of fractions* $Q(U(\mathfrak{g})/I)$ *of* $U(\mathfrak{g})/I$.

Let us remark that we don't assume a priori that I is completely prime. As usual, the operation of G on functions on G/B is the contragredient one.

We give now the connection (cf. [**Du**, 5, IV.12]) with the Duflo construction $I_{\mathfrak{g}}(?, ?)$. In the following, we use Greek letters $\varphi, \varphi_1, \ldots$ for linear forms having a solvable polarization and Latin letters $f, f_1, \ldots$ for linear forms of unipotent type.

Let $\mathfrak{g}$ be an algebraic Lie algebra and let $\varphi \in \mathfrak{g}^*$. Let $\mathfrak{c} = \mathfrak{c}_\varphi$ be the canonical subalgebra of φ. We put $\mathfrak{v} := {}^u\mathfrak{c}$ and $\delta := \varphi_{|\mathfrak{v}}$. Let $f \in \mathfrak{g}^*_{ut}$ be a form of unipotent type associated with φ, i.e., $f_{|\mathfrak{v}} = \delta$. This implies $\mathfrak{c}_f = \mathfrak{c}$ (see 8.6) and we remember that $f' \in \mathfrak{g}^*_{ut}$ and $f'_{|\mathfrak{v}} = \delta$ imply $f' \in \Gamma f$ (9.8). Let $\mathfrak{r}$ be a reductive factor of $\mathfrak{c}(\delta)$. Then by a slight modification we get from [**Du**, 5, Prop. I.28]:

15.7. PROPOSITION. *The following are equivalent.*
i) φ *admits a solvable polarization.*
ii) $\varphi_{|\mathfrak{r}}$ *admits a solvable polarization, i.e.,* $\varphi_{|\mathfrak{r}}$ *is regular.*
iii) $\varphi_{|\mathfrak{g}(f)}$ *admits a solvable polarization.*

Let us now suppose that we are in this situation, i.e., that $\varphi \in \mathfrak{g}^*_{\mathrm{sp}}$. Let us denote $\varphi_{|\mathfrak{g}(f)}$ by μ and let $\xi := J_{\mathfrak{g}(f)}(\mu)$. With this proposition in mind, we can now give the following relation given by Duflo (cf. [**Du**, 5, IV.12]).

15.8. PROPOSITION. *Let* $\varphi \in \mathfrak{g}^*_{\mathrm{sp}}$. *Let* $f \in \mathfrak{g}^*_{ut}$ *be a form of unipotent type associated with* φ. *Let* $\mu := \varphi_{|\mathfrak{g}(f)}$. *Then we have*

$$J_{\mathfrak{g}}(\varphi) = I_{\mathfrak{g}}(f, J_{\mathfrak{g}(f)}(\mu)).$$

In particular, we see that the second Duflo parameter of a Duflo ideal is itself a Duflo ideal.

We know already (15.5) that any primitive ideal contains a Duflo ideal, but now we are able to construct a canonical one. From the two preceding propositions it is easy to deduce:

15.9. COROLLARY. *Let* $f \in \mathfrak{g}^*_{ut}$ *and let* $\xi \in Z(f)$ *be primitive. Then* $I_{\mathfrak{g}}(f, \xi)$ *is a Duflo ideal if and only if* ξ *is a Duflo ideal.*

If $f \in \mathfrak{g}^*$ then the primitive ideals which are minimal among the primitive ideals of $Z(f)$ are just the Duflo ideals belonging to $Z(f)$. Hence if $\xi \in Z(f)$, let us denote by $\xi_{\min}$ the unique Duflo ideal of $Z(f)$ contained in ξ. This allows now the following definition.

15.10. DEFINITION. If $I \in \operatorname{Prim} U(\mathfrak{g})$ then let $I_{\min} := I_{\mathfrak{g}}(f, \xi_{\min})$ where (f, ξ) is any couple of Duflo parameters of I (i.e., $I = I_{\mathfrak{g}}(f, \xi)$).

Since any two couples of Duflo parameters are Γ-conjugate (§14), $I_{\min}$ is well defined and we know that $I_{\min}$ is a Duflo ideal. We have $I = I_{\min}$ if and only if I is a Duflo ideal.

15.11. DEFINITION. Let $X := \operatorname{Prim} U(\mathfrak{g})$. If J is a Duflo ideal, we put $X_J := \{I \in X | I_{\min} = J\}$.

Hence we get the following decomposition of $\operatorname{Prim} U(\mathfrak{g})$.

15.12. THEOREM. *Let* $\mathfrak{g}$ *be an algebraic Lie algebra. Then the primitive spectrum* $X = \operatorname{Prim} U(\mathfrak{g})$ *is the disjoint union of the finite sets* X_J, X_J *running through all primitive ideals of Duflo.*

15.13. REMARK. By working with solvable radicals instead of unipotent radicals, the classification theory of primitive ideals as well as Theorem 15.12 can be extended to the nonalgebraic case.

REFERENCES

[B] R. J. Blattner, *Induced and produced representations of Lie algebras*, Trans. Amer. Math. Soc. **144** (1969), 457–474.

[B-G-R] W. Borho, P. Gabriel, R. Rentschler, *Primideale in Einhüllenden auflösbarer Lie-Algebren*, Lecture Notes in Mathematics No. 357, Springer-Verlag, Berlin-Heidelberg-New York, 1973.

[C, 1] N. Conze, *Algèbres d'opérateurs différentiels et quotients des algèbres enveloppantes*, Bull. Soc. Math. France **102** (1974), 379–415.

[C, 2] N. Conze, *Espace des idéaux primitifs de l'algèbre enveloppante d'une algèbre de Lie nilpotente*, J. Algebra **34** (1975), 444–450.

[C-Du] N. Conze, M. Duflo, *Sur l'algèbre enveloppante d'une algèbre de Lie résoluble*, Bull. Sci. Math. **96** (1972), 339–351.

[Di, 1] J. Dixmier, *Représentations irréductibles des algèbres de Lie résolubles*, J. Math. Pures Appl. **45** (1966), 1–66.

[Di, 2] J. Dixmier, *Idéaux primitifs dans l'algèbre enveloppante d'une algèbre de Lie semi-simple complexe*, C. R. Acad. Sci. **271**, série A (1970), 134–136.

[Di, 3] J. Dixmier, *Sur les représentations induites des algèbres de Lie*, J. Math. Pures Appl. **50** (1971), 1–24.

[Di, 4] J. Dixmier, *Polarisations dans les algèbres de Lie*, Ann. Sci. École Norm. Sup. (4) (1971), 321–336.

[Di, 5] J. Dixmier, *Algèbres enveloppantes*, Cahiers Scientifiques XXXVII, Gauthier-Villars, Paris, 1974.

[Di, 6] J. Dixmier, *Idéaux primitifs dans les algèbres enveloppantes*, J. Algebra **48** (1977), 96–112.

[Du, 1] M. Duflo, *Sur les extensions des représentations irréductibles des groupes de Lie nilpotents*, Ann. Sci. École Norm. Sup. 5 (1972), 71–120.

[Du, 2] M. Duflo, *Certain algèbres de type fini sont des algèbres de Jacobson*, J. Algebra **27** (1973), 358–365.

[Du, 3] M. Duflo, *Construction of primitive ideals in an enveloping algebra*, Proceedings of the Summer School of Group Representations, Bolya; Janos Mathematical Society (Budapest, 1971), edited by I. M. Gelfand, Adam Hilger, London and Akad. Kiadó, Budapest, 1975.

[Du, 4] M. Duflo, *Sur la classification des idéaux primitifs dans l'algèbre enveloppante d'une algèbre de Lie semi-simple*, Ann. of Math. **105** (1977), 107–120.

[Du, 5] M. Duflo, *Théorie de Mackay pour les groupes de Lie algébriques*, Acta Math. **149** (1982), 153–213.

[Du, 6] M. Duflo, *Sur les idéaux induits dans les algèbres enveloppantes*, Invent. Math. **67** (1982), 385–393.

[Ja] J. C. Jantzen, *Einhüllende Algebren halbeinfacher Lie-Algebren*, Springer-Verlag, Berlin-Heidelberg-New York-Tokyo, 1983.

[Jo] A. Joseph, *Primitive ideals in enveloping algebras*, Proceedings of the International Congress of Mathematicians, August 16-24, 1983, Warzawa, pp. 403–414.

[M] C. Moeglin, *Idéaux primitifs des algèbres enveloppantes*, J. Math. Pures Appl. **59** (1980), 265–336.

[M-R, 1] C. Moeglin, R. Rentschler, *Orbites d'un groupe algébrique dans l'espace des idéaux rationnels d'une algèbre enveloppante*, Bull. Soc. Math. France **109** (1981), 403–426.

[M-R, 2] C. Moeglin, R. Rentschler, *Sur la classification des idéaux primitifs des algèbres enveloppantes*, Bull. Soc. Math. France **112** (1984), 3–40.

[M-R, 3] C. Moeglin, R. Rentschler, *Sous-corps ad-stables des anneaux de fractions des quotients des algèbres enveloppantes, espaces homogènes et induction*, de Mackey, J. Funct. Anal. **69** (1986), 307–396.

[M-R, 4] C. Moeglin, R. Rentschler, *Idéaux G-rationnels, rang de Goldie*, preprint, Paris, 1986.

[P] A. N. Panov, *The Dixmier map*, Funktsional. Anal. i Prilozhen. **14**, No. 1 (1980), 79–80 (English translation: Functional Anal. Appl. **14**, No. 1 (1980), 65–66).

[R] R. Rentschler, *L'injectivité de l'application de Dixmier pour les algèbres de Lie résolubles*, Invent. Math. **23** (1974), 49–71.

[T, 1] P. Tauvel, *Sur la bicontinuité de l'application de Dixmier pour les algèbres de Lie résolubles*, Ann. Fac. Sci. Toulouse Math. 4 (1982), 291–308.

[T, 2] P. Tauvel, *Modules induits sur les algèbres enveloppantes*, Bull. Sci. Math. (2) **109** (1985), 379–397.

[V] M. Vergne, *Construction de sous-algèbres subordonnées à un élément du dual d'une algèbre de Lie résoluble*, C. R. Acad. Sci. **270**, série A (1970), 173–175.

Filtered Noetherian Rings

JAN-ERIK BJÖRK

Introduction. We are going to study noncommutative rings equipped with filtrations given by increasing sequences of additive subgroups $\Sigma_{v-1} \subset \Sigma_v \subset \Sigma_{v+1}$.

If R is such a filtered ring we get its associated graded ring $\mathrm{gr}(R)$. If M is an R-module (left or right) we can consider filtrations on it and get the corresponding graded $\mathrm{gr}(R)$-module. The subsequent material studies the interplay between R-modules M and their associated $\mathrm{gr}(R)$-modules $\mathrm{gr}_\Gamma(M)$ where Γ is some filtration on M.

Of course, this only leads to substantial results under additional conditions. First we assume that both R and $\mathrm{gr}(R)$ are Noetherian rings.

Since they need not be commutative it means that they are both left and right Noetherian. If M is a finitely generated R-module one finds a distinguished class of *good filtrations* on M. This is explained in §2 of Part I.

In Part I we then develop a theory where both R and $\mathrm{gr}(R)$ are Gorenstein rings. Here the *noncommutative Gorenstein condition* is introduced.

We refer to the summary of Part I for more comments. In Part II we study the case when $\mathrm{gr}(R)$ is commutative and Noetherian and prove that characteristic ideals of finitely generated R-modules are stable under the Poisson product on $\mathrm{gr}(R)$.

In §3 we offer examples of the preceding material. In particular, we discuss the Weyl algebra $A_n(\mathbb{C})$, and some of the results in §3 have been the inspiration for the general study in Part I and Part II.

Final Remarks. We refer to the summary of each part for a more detailed presentation of their content. In Part I and Part II we offer rather detailed proofs. However, we assume that the reader knows the basic constructions of spectral sequences attached to filtered complexes and also some basic results from ring theory and homological algebra. In Part III we mention various results which go beyond the present material since they require methods which go far beyond those in Part I.

I. Filtered Gorenstein Rings

Summary. In §1 we consider a ring R which is both left and right Noetherian. Using projective resolutions of finitely generated R-modules and the trivial fact that finitely generated projective modules are reflexive one gets the *biduality formula* $M = \mathbb{R}\operatorname{Hom}_R(\mathbb{R}\operatorname{Hom}_R(M,R),R)$ which is an equality in the derived category of complexes of R-modules. We explain this in detail, so the reader does not have to know the whole story about derived categories. Instead this equality (or rather isomorphism) is made clear by projective resolutions which lead to the so-called *bidualizing complex* which we can attach to any finitely generated R-module.

The bidualizing complex is found as the simple diagonal complex of a double complex and hence it can be equipped with two filtrations, the first and the second. The first degenerates and gives the biduality formula while the second in general has higher terms which in degree 2 are expressed by abelian groups of the form $\operatorname{Ext}_R^v(\operatorname{Ext}_R^j(M,R),R)$.

In particular, the second filtration induces a filtration on any finitely generated R-module in a canonical way. We call this the $\mathcal{B}$-filtration.

This $\mathcal{B}$-filtration is used to measure the size of finitely generated R-modules. In order to get more precise information we add assumptions. First we impose the condition that R has a *finite self-injective dimension*, i.e., that $\operatorname{Ext}_R^v(M,R) = 0$ for all R-modules M (left or right) and all $v > \mu = \operatorname{inj.dim}(R)$.

In this case, the $\mathcal{B}$-filtrations on modules have at most μ strict inclusions. So when M is a finitely generated R-module we get a filtration $\mathcal{B}_0(M) \subset \mathcal{B}_1(M) \subset \cdots \subset \mathcal{B}_\mu(M) = M$ and this suggests that we define $\delta(M)$ as the unique smallest integer for which $M = \mathcal{B}_{\delta(M)}(M)$ holds.

The *Gorenstein condition* is introduced in §1.16 and it is basic for Part I. It is a condition which enables us to control the nondegenerated spectral sequence arising from the second filtration on the bidualizing complexes of finitely generated R-modules. In particular, we discover that it gives the equality $\delta(M)+j(M) = \mu$ for any finitely generated R-module M. Here $j(M)$ is the smallest integer such that $\operatorname{Ext}_R^{j(M)}(M,R) \neq 0$.

Filtered rings are discussed in §2. We will only study the case when R is a filtered ring such that both R and $\operatorname{gr}(R)$ are left and right Noetherian. In addition to this we need one more condition on the filtration, namely the so-called *comparison condition* which is defined in §2.5.

We remark that this condition can hold in quite general situations and a stronger condition is mentioned in §2.11, where it is required that a certain family of additive subgroups of R are closed with respect to the filtration. It is called the *closure condition* and we show that it implies the comparison condition.

We refer to Part III for examples of filtered rings where the comparison condition holds but not the closure condition.

In §3 we have only recalled some standard constructions which are used in the proofs later on. A notable point is that the comparison condition implies that spectral sequences attached to filtered complexes of the form $0 \to \mathrm{Hom}_R(F_0, N) \to \mathrm{Hom}_R(F_1, N) \to \cdots$ converge. Here N is equipped with some good filtration and $F.$ is a "graded free resolution" of some filtered R-module M.

The case when $N = R$ is, in particular, studied and in §4 we reach the main results of Part I. For example, we prove that if $\mathrm{gr}(R)$ is Gorenstein so is R under the assumption that the filtration satisfies the closure condition. In addition, we get a remarkable inequality in Theorem 4.4 under the weaker assumption that the comparison condition holds. We refer to Part III for an application of this result.

The case when $\mathrm{gr}(R)$ is *commutative* is treated in the final sections. Here we find the characteristic ideal $J(M)$ of any finitely generated R-module M and we study its prime decomposition in some detail. For example, we find an interplay between the $\mathcal{B}$-filtration of M and the presence of minimal prime divisors $\mathcal{P}$ of $J(M)$ where $\mathrm{inj.dim}(\mathrm{gr}(R)_{\mathcal{P}})$ are prescribed.

Final Remarks. The material in §§1–3 is included in a far more general theory in [9]. There the Gorenstein condition is contained in quite general conditions concerned with "pseudo-duality". We remark that Auslander suggested the Gorenstein condition in the case of noncommutative rings. The preservation of the Gorenstein condition as announced in Theorem 4.1 was announced by J.-E. Roos in [27]. So here we have supplied detailed proofs and the remarkable inequality in Theorem 4.4 is new. Here we use a "backward induction" which shows how efficient a systematic study of higher terms in a spectral sequence can be.

The material in §§5 and 6 is taken from lectures by O. Gabber which also have been revised by Levasseur in [11] and we also refer to his thesis [23] which contains closely related results. Let us also mention [4: Chapter 2] where the special case $\mathrm{gr}(R)$ a commutative Noetherian regular ring is treated. A notable fact is that *a regular commutative Noetherian ring always is Gorenstein.* See for example [4: pp. 65–69] for a detailed proof.

1. Noetherian Gorenstein rings.

Let R be an associative ring with a multiplicative unit. All R-modules (left or right) are unitary. We do not assume that R is commutative and when we say that R is Noetherian it means that it is both left and right Noetherian.

1.1. *The equality* $M = \mathbb{R}\mathrm{Hom}(\mathbb{R}\mathrm{Hom}(M, R), R)$. Assume that R is Noetherian. If M is a finitely generated left (or right) R-module then we get the equality above in the derived category of complexes of R-modules. This follows because finitely generated projective R-modules are reflexive and from the existence of projective resolutions of M, where we only need finitely generated projectives to get such a resolution.

However, we give a rather explicit description of this equality because it will be used to get more refined information later on. So let M be a finitely generated R-module and choose a projective resolution $\cdots \to P_1 \to P_0 \to M \to 0$ where $P_0, P_1, \dots$ are finitely generated projective R-modules. In general this resolution is infinite, i.e., it never stops unless the projective dimension of M is finite.

Now $\mathbb{R}\operatorname{Hom}(M, R)$ is realized by the complex $0 \to \hat{P}_0 \to \hat{P}_1 \to \cdots$ where $\hat{P}_v = \operatorname{Hom}(P_v, R)$ are finitely generated and projective *right* R-modules. To be precise, the right R-module structures follow from the obvious bimodule structure on the ring R itself.

To get $\mathbb{R}\operatorname{Hom}(\mathbb{R}\operatorname{Hom}(M, R), R)$ we need a projective resolution of the complex $\hat{P}^{\cdot}$. It consists of a double complex $Q_{\cdot\cdot}$ where Q_{vk} are finitely generated and projective right R-modules which give rise to a commutative diagram:

$$\begin{array}{ccccccc}
Q_{10} & \to & Q_{11} & \to & Q_{12} & \to \\
\downarrow & & \downarrow & & \downarrow & \\
Q_{00} & \to & Q_{01} & \to & Q_{02} & \to \\
\downarrow & & \downarrow & & \downarrow & \\
\hat{P}_0 & \to & \hat{P}_1 & \to & \hat{P}_2 & \to
\end{array}$$

Each column $0 \leftarrow \hat{P}_k \leftarrow Q_{0k} \leftarrow Q_{1k} \leftarrow \cdots$ is exact. In the row complexes $0 \to Q_{j0} \to Q_{j1} \to \cdots$ the cohomology groups $\mathcal{H}^k(Q_{j\cdot})$ are projective right R-modules and for each k we get an induced exact sequence

$$0 \leftarrow \mathcal{H}^k(\hat{P}^{\cdot}) \leftarrow \mathcal{H}^k(Q_{0\cdot}) \leftarrow \mathcal{H}^k(Q_{1\cdot})$$

where we observe that $\mathcal{H}^k(\hat{P}^{\cdot}) = \operatorname{Ext}^k(M, R)$.

Remark. The construction of $Q_{\cdot\cdot}$ is standard. See [13] or [4: pp. 59–61] for details if necessary.

Now $\mathbb{R}\operatorname{Hom}(\mathbb{R}\operatorname{Hom}(M, R), R)$ is realized by the simple diagonal complex associated with the double complex $\hat{Q}_{\cdot\cdot}$. To be precise, $\hat{Q}_{\cdot\cdot}$ is arranged as a double complex which is concentrated in the second quadrant:

$$\begin{array}{cccccc}
 & & & \hat{Q}_{20} & \to & 0 \\
 & \uparrow & & \uparrow & & \\
\to & \hat{Q}_{11} & \to & \hat{Q}_{10} & \to & 0 \\
 & \uparrow & & \uparrow & & \\
\to & \hat{Q}_{01} & \to & \hat{Q}_{00} & \to & 0
\end{array}$$

Now $\Delta^k = \oplus \hat{Q}_{k+v,v}$ give $\Delta^{k-1} \to \Delta^k \to \Delta^{k+1}$ and we can announce the equality in 1.1 as follows.

1.2. PROPOSITION. *$\Delta^{\cdot}$ is exact in all degrees except zero where $\mathcal{H}^0(\Delta^{\cdot})$ is isomorphic to M.*

Proof. Consider the *first filtration* on $\Delta^{\cdot}$ where we begin to compute the cohomology along the columns in $\hat{Q}_{\cdot\cdot}$. Since $0 \leftarrow \hat{P}_k \leftarrow Q_{0k} \leftarrow Q_{1k}$ are exact and all modules are projective, it follows that $0 \to \hat{\hat{P}}_k \to \hat{Q}_{0k} \to \hat{Q}_{1k} \to \cdots$ are exact for all k. This shows that the first term of the spectral sequence is reduced to a single row $\cdots \to P_{-k-1} \to P_{-k} \to P_{-k+1} \to \cdots$ where $P_{-k} \cong \hat{\hat{P}}_{-k}$ are used. Passing to the second term we see that only an isomorphic copy of M remains in position $(0,0)$ and Proposition 1.2 follows.

REMARK. See [9] or [13] or [4: pp. 52–54] for standard facts about double complexes and their associated spectral sequences arising from the first and second filtration, respectively.

1.3. *The nondegenerated spectral sequence.* If we use instead the second filtration on $\Delta^{\cdot}$ where the first term arises when we compute the cohomology along the rows of $\hat{Q}_{..}$ then one has

1.4. *Formula.* The second term is a complex $E_2^{k-1} \to E_2^k \to E_2^{k+1}$ where $E_2^k = \oplus E_2^k(v)$ and here $E_2^k(v) = \operatorname{Ext}^{k-v}(\operatorname{Ext}^{-v}(M,R),R)$ for all k and v.

See, for example, [9] or [13] for this formula which uses that $\mathcal{H}^k(\hat{P}_{\cdot}) = \operatorname{Ext}^k(M,R)$ for all k and the projective resolutions $0 \leftarrow \mathcal{H}^k(\hat{P}_{\cdot}) \leftarrow \mathcal{H}^k(Q_{0\cdot}) \leftarrow \mathcal{H}^k(Q_{1\cdot}) \leftarrow \cdots$.

Remark. In the spectral sequence which is found from the second filtration on $\Delta^{\cdot}$ we know that the differentials in the complex $E_2^{\cdot}$ are homogeneous of degree -1. This means that $E_2^k(v) \to E_2^{k+1}(v-1)$ exist for all k and v and using the formula in §1.4, this means that canonically defined mappings from $\operatorname{Ext}^{k-v}(\operatorname{Ext}^{-v}(M,R),R) \to \operatorname{Ext}^{k+2-v}(\operatorname{Ext}^{-v+1}(M,R),R)$ exist for all k and v.

1.5. *The Gorenstein condition.* If we use the second filtration, we have found expressions of the second terms of the spectral sequence attached to the filtered complex $\Delta^{\cdot}$. With no further assumptions, we cannot say much about this spectral sequence. So we begin to impose more conditions on the Noetherian ring R.

1.6. *Condition.* Assume that R has a finite injective dimension, i.e., we assume that there exists an integer μ so that $\operatorname{Ext}^v(M,R) = 0$ for any R-module M (left or right) and all $v > \mu$.

In this case we see that the complex $E_2^{\cdot}$ is bounded, i.e., $E_2^k = 0$ both when $k > \mu$ and when $k < -\mu$. In fact, the table for $E_2^{\cdot}$ is even bounded and this implies that the spectral sequence converges. This means that there exists an integer r so that the higher term $E_r^{\cdot} \cong E_\infty^{\cdot}$ where E_∞^k by definition are associated graded groups of the filtered cohomology groups $\mathcal{H}^k(\Delta^{\cdot})$.

Using Proposition 1.2 we have $E_\infty^k = 0$ for all $k \neq 0$ while $E_\infty^0 = \operatorname{gr}(M)$ where the R-module M is equipped with a filtration which under Condition 1.6 has at most μ strict inclusions.

To be precise, $\mathcal{B}_0(M) \subseteq \mathcal{B}_1(M) \subseteq \cdots \subseteq \mathcal{B}_\mu(M) = M$ occurs where $\mathcal{B}_v(M)$ are R-submodules. We call $\mathcal{B}_{\cdot}(M)$ the filtration on M induced by the bidualizing complex.

1.7. *The Gorenstein condition.* Let M be a finitely generated R-module (left or right). We assume Condition 1.6 and get the $\mathcal{B}$-filtration which suggests

1.8. *Definition.* $\delta(M)$ is the unique smallest integer for which $M = \mathcal{B}_{\delta(M)}(M)$.

Next, let us consider the Ext-groups $\mathrm{Ext}^v(M,R)$. They vanish if $v > \mu$ and a notable fact is that they cannot vanish for all $0 \le v \le \mu$. This follows from the previous formulas, i.e., $\mathrm{gr}(M) = E_\infty^0$ is a *subquotient* of $E_2^0 = \oplus\,\mathrm{Ext}^{-v}(\mathrm{Ext}^{-v}(M,R),R)$.

Remark. In general, if H is an abelian group then a subquotient of H is a group of the form H_1/H_0 where $H_0 \subset H_1 \subset H$. In a spectral sequence we find that $E_{r+1}^k = \mathcal{H}^k(E_r^\cdot)$ are subquotients of E_r^k for all k and r. From Condition 1.6 we know that the spectral sequence converges so that $\mathrm{gr}(M) = E_\infty^0 \cong E_r^0$ for some large r and now E_r^0 is isomorphic to a subquotient of E_2^0.

Remark. The indices which are used to define the $\mathcal{B}$-filtration should really be put in negative order. To be precise, $E_2^0 = \oplus E_2^0(v) = \oplus\,\mathrm{Ext}^{-v}(\mathrm{Ext}^{-v}(M,R),R) \Rightarrow E_2^0(v)$ can only differ from zero when $-\mu \le v \le 0$ and passing to E_∞^0 we have used $E^0(v-\mu) = \mathcal{B}_v(M)/\mathcal{B}_{v-1}(M)$ for all integers v. This clarifies how the indices are arranged when the $\mathcal{B}$-filtration on M is introduced and it gives

1.9. *Formula.* $\mathcal{B}_v(M)/\mathcal{B}_{v-1}(M)$ are subquotients of $E_2^0(v-\mu)$.

Example. $\mathcal{B}_0(M)$ is a subquotient of $\mathrm{Ext}^\mu(\mathrm{Ext}^\mu(M,R),R)$.

1.10. *Definition.* If M is a finitely generated R-module (not $\equiv 0$) then $j(M)$ is the unique smallest integer for which $\mathrm{Ext}^{j(M)}(M,R) \neq 0$.

The preceding discussion shows that $j(M)$ exists and Condition 1.6 gives $0 \le j(M) \le \mu$. Using Formula 1.9 we get

1.11. *Formula.* $j(M) + \delta(M) \le \mu$.

Proof. Since $\mathcal{B}_v(M)/\mathcal{B}_{v-1}(M)$ are subquotients of $\mathrm{Ext}^{\mu-v}(\mathrm{Ext}^{\mu-v}(M,R),R)$, the definition of $j(M)$ gives $\mathcal{B}_v(M)/\mathcal{B}_{v-1}(M) = 0$ when $\mu - v < j(M)$ and starting from $M = \mathcal{B}_\mu(M)$ we get $M = \mathcal{B}_{\mu-j(M)}(M)$ and hence $\delta(M)$ is $\le \mu - j(M)$.

1.12. *The equality* $j(M) + \delta(M) = \mu$. If we only assume Condition 1.6 then this equality fails in general. Various counterexamples occur in [9] where we, in particular, refer to [9: p. 105].

So we try to impose more conditions in order to ensure this equality for all finitely generated R-modules. First we can try

1.13. *Condition.* $\mathrm{Ext}^v(\mathrm{Ext}^j(M,R),R) = 0$ for all pairs $v < j$ and any finitely generated R-module M.

Assuming this, we discover that $E_2^k = 0$ for all $k < 0$. This simplifies the passage to the limit of the spectral sequence. In particular, no coboundaries occur in degree zero. For example, E_3^0 is a submodule of E_2^0 and so on. We find that $E_\infty^0 \subset E_2^0$ which means that Condition 1.13 implies

1.14. *Formula.* $\mathcal{B}_v(M)/\mathcal{B}_{v-1}(M)$ are submodules of $E_2^0(v-\mu)$ for all v.

Unfortunately, Condition 1.13 is not enough to get $j(M) + \delta(M) = \mu$ in general. See again [9: p 105] for an explicit counterexample.

1.15. *The Gorenstein condition.* Observe that the definition of the j-numbers means that Condition 1.13 asserts that $j(\mathrm{Ext}^m(M,R)) \ge m$ for all m and any finitely generated R-module M. A stronger condition is

1.16. *Condition.* For any finitely generated R-module M and any integer m and any submodule N of $\mathrm{Ext}^m(M,R)$, $j(N) \geq m$.

We call this the Gorenstein condition and, when it holds, we can prove the equality $j(M) + \delta(M) = \mu$.

Proof. Take a finitely generated R-module M and, for a given $0 \leq v \leq \mu$, consider $N = \mathrm{Ext}^{\mu-v}(M,R)$. Now $\mathcal{B}_v(M)/\mathcal{B}_{v-1}(M)$ is a submodule of $\mathrm{Ext}^{\mu-v}(\mathrm{Ext}^{\mu-v}(M,R),R) = \mathrm{Ext}^{\mu-v}(N,R)$ and Condition 1.16 implies that $j(\mathcal{B}_v(M)/\mathcal{B}_{v-1}(M)) \geq \mu - v$.

Using an induction over v, starting from $v = 0$ we get $j(\mathcal{B}_v(M)) \geq \mu - v$ for all v. In fact, just use the long exact sequences of Ext-groups which arise from $0 \to \mathcal{B}_{v-1}(M) \to \mathcal{B}_v(M) \to \mathcal{B}_v(M)/\mathcal{B}_{v-1}(M) \to 0$ when R is used as second factor.

Finally, $M = \mathcal{B}_{\delta(M)}(M) \Rightarrow j(M) \geq \mu - \delta(M)$ and hence $j(M) + \delta(M) \geq \mu$ which, together with Formula 1.11, proves the equality $j(M) + \delta(M) = \mu$.

Summing up, we say that R is a Noetherian Gorenstein ring when R has a finite injective dimension μ and Condition 1.16 holds. We have then seen that $j(M) + \delta(M) = \mu$ holds for any finitely generated R-module M. This equality can be used to clarify the size of finitely generated R-modules.

1.17. *Purity conditions.* If M is a finitely generated R-module we say that it has a pure δ-dimension if $\mathcal{B}_{\delta(M)-1}(M) = 0$, i.e., the $\mathcal{B}$-filtration degenerates to a single term $= M$. We want to analyze this concept of purity and begin with a rather technical result.

1.18. PROPOSITION. *$\mathrm{Ext}^{j(M)}(M,R)$ has the pure δ-dimension $\delta(M)$ for any finitely generated R-module M.*

Proof. Consider the bidualizing complex of M. We find the second term, which is a complex $E_2^{\cdot}$ where $E_2^k(v) = \mathrm{Ext}^{k-v}(\mathrm{Ext}^{-v}(M,R),R)$ and the differentials send $E_2^k(v) \to E_2^{k+1}(v-1)$.

The definition of $j(M)$ implies $E_2^k(-j(M)+v) = 0$ for all $v \geq 1$ and all k. Recall that the differentials in $E_2^{\cdot}$ are homogeneous of degree -1. In particular, we get $E_2^{k-1}(-j(M)+1) \to E_2^k(-j(M)) \to E_2^{k+1}(-j(M)-1)$, which shows that no coboundaries occur in position $(k, -j(M))$ since $E_2^{k-1}(-j(M)+1) = 0$ for all k. This gives

SUBLEMMA 1. *For any integer k it follows that $0 \to E_3^k(-j(M)) \to E_2^k(-j(M)) \to S_2^k \to 0$ is an exact sequence with $S_2^k \subseteq E_2^{k+1}(-j(M)-1)$.*

In $E_3^{\cdot}$ we also get $E_3^k(-j(M)+v) = 0$ for all k and all $v \geq 1$ and if we recall that the differentials in $E_3^{\cdot}$ are homogeneous of degree -2 we discover that no coboundaries occur in position $(k, -j(M))$ and this gives

SUBLEMMA 2. *$0 \to E_4^k(-j(M)) \to E_3^k(-j(M)) \to S_3^k \to 0$ are exact with $S_3^k \subseteq E_3^{k+1}(-j(M)-2)$ for all integers k.*

The same phenomenon occurs in general, i.e., to each $r \geq 2$ we have $0 \to E_{r+1}^k(-j(M)) \to E_r^k(-j(M)) \to S_r^k \to 0$ with $S_r^k \subseteq E_r^{k+1}(-j(M)-r+1)$.

In order to use these exact sequences we shall need

SUBLEMMA 3. *$j(S_r^k) \geq k + j(M) + r$ for all $r \geq 2$.*

Proof. S_r^k is a subquotient of

$$E_2^{k+1}(-j(M) - r + 1) = \operatorname{Ext}^{k+r+j(M)}(\operatorname{Ext}^{j(M)+r-1}(M, R), R)$$

and the last module has j-number $\geq k + r + j(M)$ and so has its subquotient S_r^k by the Gorenstein condition.

Let us now fix some $k \geq 1$. We know that the spectral sequence converges and here $E_\infty^k = \operatorname{gr}(\mathcal{H}^k(\Delta^\cdot)) = 0$ since $k \neq 0$. Now $0 = E_\infty^k = E_r^k$ for some large r. Starting from this r we use Sublemma 3 and a (decreasing) induction using the exact sequences $0 \to E_{v+1}^k(-j(M)) \to E_v^k(-j(M)) \to S_v^k \to 0$ to get

SUBLEMMA 4. *$j(E_2^k(-j(M))) \geq k + j(M) + 2$.*

Final part of the proof. We have $E_2^k(-j(M)) = \operatorname{Ext}^{k+j(M)}(\operatorname{Ext}^{j(M)}(M, R), R)$ $= \operatorname{Ext}^{k+j(M)}(N, R)$ with $N = \operatorname{Ext}^{j(M)}(M, R)$. Sublemma 4 and the definition of the j-number gives $\operatorname{Ext}^{k+j(M)}(\operatorname{Ext}^{k+j(M)}(N, R), R) = 0$. This holds for any $k \geq 1$. Using formula 1.14 it gives $\mathcal{B}_v(N)/\mathcal{B}_{v-1}(N) = 0$ for all $0 \leq v < \delta(M) = \mu - j(M)$. Finally, $N = \mathcal{B}_{\delta(M)}(N)$ holds because the j-number of $N = \operatorname{Ext}^{j(M)}(M, R)$ is at least $j(M)$ which, together with the equality $j(N) + \delta(N) = \mu$, implies $\delta(N) \leq \delta(M)$. Since we have proved that $\mathcal{B}_v(N)/\mathcal{B}_{v-1}(N) = 0$ for all $v < \delta(M)$ we conclude that N has pure δ-dimension.

Remark. Observe that Sublemma 4 above gives even more, namely that $\operatorname{Ext}^{k+j(M)+1}(\operatorname{Ext}^{k+j(M)}(\operatorname{Ext}^{j(M)}(M, R), R), R) = 0$ for all $k \geq 1$. However, this refined information will not be used so it is only mentioned.

1.19. *Applications.* Using Proposition 1.18, we get

1.20. COROLLARY. *$\operatorname{Ext}^k(\operatorname{Ext}^k(M, R), R)$ is either zero or of pure δ-dimension $\mu - k$ for all $0 \leq k \leq \mu$ and any finitely generated R-module M.*

Proof. Take $N = \operatorname{Ext}^k(M, R)$, which has $j(N) \geq k$ by the Gorenstein condition. If $\operatorname{Ext}^k(\operatorname{Ext}^k(M, R), R) \neq 0$, then $j(N) = k$ and $\operatorname{Ext}^k(N, R) = \operatorname{Ext}^k(\operatorname{Ext}^k(M, R), R)$ has pure δ-dimension $= \mu - k$.

Next, considering some R-module M and using Formula 1.14, we see that if $\operatorname{Ext}^{\mu-v}(\operatorname{Ext}^{\mu-v}(M, R), R) = 0$ for some v, then $\mathcal{B}_v(M)/\mathcal{B}_{v-1}(M) = 0$. It turns out that the converse is true also, i.e., one has

1.21. PROPOSITION. *If $\mathcal{B}_v(M)/\mathcal{B}_{v-1}(M) = 0$, then*

$$\operatorname{Ext}^{\mu-v}(\operatorname{Ext}^{\mu-v}(M, R), R) = 0.$$

Proof. In the spectral sequence, which as usual is attached to the second filtration of the bidualizing complex of M, we get exact sequences

$$0 \to E_{r+1}^0(v - \mu) \to E_r^0(v - \mu) \to S_r \to 0 \quad \text{for all } r \geq 2.$$

Here

$$S_r \subseteq E_r^1(v-\mu-r+1) = \text{ a subquotient of } E_2^1(v-\mu-r+1)$$
$$= \operatorname{Ext}^{\mu-v+r}(\operatorname{Ext}^{\mu+r-v-1}(M,R),R)$$

and this R-module has j-number $\geq \mu - v + r \geq \mu - v + 2$ for all $r \geq 2$. Hence $j(S_r) \geq \mu - v + 2$ for all r.

Now $\mathcal{B}_v(M)/\mathcal{B}_{v-1}(M) = E_\infty^0(v-\mu) \cong E_r^0(v-\mu)$ when r is large. So if $\mathcal{B}_v(M)/\mathcal{B}_{v-1}(M) = 0$, then $E_r^0(v-\mu) = 0$ for some large r and by a backward induction we arrive at $j(E_2^0(v-\mu)) \geq \mu - v + 2$. Now we have $E_2^0(v-\mu) = \operatorname{Ext}^{\mu-v}(\operatorname{Ext}^{\mu-v}(M,R),R)$ and if it is $\neq 0$ its δ-dimension equals v by Corollary 1.20, so its j-number $= \mu - v$ and we find a contradiction.

1.22. COROLLARY. *The R-module M has a pure δ-dimension if and only if* $\operatorname{Ext}^v(\operatorname{Ext}^v(M,R),R) = 0$ *for all* $v > j(M)$.

1.23. *A useful imbedding.* Let M be an R-module with a pure δ-dimension. Hence $M = \mathcal{B}_{\delta(M)}(M)/\mathcal{B}_{\delta(M)-1}(M)$ and the last module is a submodule of $M^\times = \operatorname{Ext}^{j(M)}(\operatorname{Ext}^{j(M)}(M,R),R)$ which also has a pure δ-dimension by Corollary 1.20. Also, the proof of Proposition 1.21 shows that

$$0 \to M \to M^\times \to M^\times/M \to 0 \quad \text{is exact with} \quad \delta(M^\times/M) \leq \delta(M) - 2.$$

Suppose now that N is some R-module of pure δ-dimension which contains M and satisfies $\delta(N/M) \leq \delta(M) - 2$. Then we find that N is isomorphic to a submodule of $M^\times$.

Proof. Using a long exact sequence of Ext, applied to $0 \to M \to N \to N/M \to 0$ and the fact that $j(N/M) \geq j(M) + 2 \Rightarrow \operatorname{Ext}^{j(M)+1}(N/M,R) = 0$, we get $\operatorname{Ext}^{j(M)}(M,R) \cong \operatorname{Ext}^{j(M)}(N,R)$ and hence $M^\times \cong N^\times$. Also, $N \subset N^\times$ since N has a pure δ-dimension and we get $N \subset M^\times$. So we can say that $M^\times$ is the largest module which contains M, under the condition that it has pure δ-dimension and satisfies $\delta(M^\times/M) \leq \delta(M) - 2$.

1.24. *Holonomic R-modules.* Let R be a Noetherian Gorenstein ring. A finitely generated R-module M satisfying $\delta(M) = 0$ is called *holonomic*. If μ is the injective dimension of R then we have seen that $\delta(M) = 0 \Rightarrow j(M) = \mu$ and this means that $\operatorname{Ext}_R^v(M,R) = 0$ for all $0 \leq v < \mu$.

1.25. *The functor $M \to \tilde{M}$.* If M is a holonomic left R-module we put $\tilde{M} = \operatorname{Ext}_R^\mu(M,R)$, which is a holonomic right R-module. Since M and $\mathbb{R}\operatorname{Hom}(\mathbb{R}\operatorname{Hom}(M,R),R)$ are isomorphic and M is holonomic we see that M is isomorphic to $\operatorname{Ext}^\mu(\operatorname{Ext}^\mu(M,R),R) = \operatorname{Ext}^\mu(\tilde{M},R)$. This gives

1.26. PROPOSITION. *$M \to \tilde{M}$ is an exact contravariant functor from the category of holonomic left R-modules to the category of holonomic right R-modules.*

1.27. *A useful consequence.* A holonomic R-module has a finite length as an R-module, i.e., it is both Noetherian and Artinian. To see this we use the exact functor above and observe that if M is holonomic and if $M \supset M_1 \supset M_2 \supset \cdots$

is decreasing then $(M/M_1)^\sim \subset (M/M_2)^\sim \subset \cdots$ increases in the Noetherian module M. This proves that the R-module M is Artinian.

2. Filtered rings. By a filtration on a ring R we mean an increasing sequence of additive subgroups $\Sigma_{v-1} \subset \Sigma_v \subset \Sigma_{v+1}$ satisfying $\bigcup \Sigma_v = R$ and $\bigcap \Sigma_v = \{0\}$ and the inclusions $\Sigma_v \Sigma_k \subset \Sigma_{v+k}$ hold for all pairs of integers v and k.

Remark. We assume that R has an identity which belongs to Σ_0. Observe that we allow $\Sigma_v \neq 0$ for all v. If $\Sigma_v = 0$ when $v < 0$ one says that the filtration is *discrete*. Observe that we can have filtrations where $\Sigma_v = \Sigma_0$ for all $v > 0$ while $\Sigma_0 \supset \Sigma_{-1} \supset \Sigma_{-2} \supset \cdots$ is strictly decreasing. The last example occurs if we take a local ring and use the $\mathfrak{m}$-adic filtration with $\Sigma_{-v} = \mathfrak{m}^v$ for all $v \geq 1$.

Let R be a filtered ring. A filtration on a left R-module M consists of an increasing sequence $\Gamma_{v-1} \subset \Gamma_v \subset \Gamma_{v+1}$ satisfying $\Sigma_k \Gamma_v \subset \Gamma_{k+v}$ for all k and v.

Remark. We do not assume that $\bigcup \Gamma_v = M$ or that $\bigcap \Gamma_v = \{0\}$. If $\bigcup \Gamma_v = M$ one says that the filtration is *exhaustive* and if $\bigcap \Gamma_v = \{0\}$ one says that the filtration is *separated.*

2.1. *Good filtrations.* If M is a finitely generated R-module we find a family of special filtrations on M as follows: Let $u_1, \ldots, u_s$ be some finite set of elements in M which are generators, i.e., $M = Ru_1 + \cdots + Ru_s$ and let $m_1, \ldots, m_s$ be integers. Put $\Gamma_v = \Sigma_{v-m_1} u_1 + \cdots + \Sigma_{v-m_s} u_s$ for all v. Then $\{\Gamma_v\}$ is a filtration on M. The family of all such filtrations, which depend on the chosen generators and the integers $m_1, \ldots, m_s$ as above, is denoted by $\mathcal{F}(M)$. Any filtration in the family $\mathcal{F}(M)$ is a good filtration.

2.2. *The $\mathcal{F}$-topology.* Let M be a finitely generated R-module. If $\Gamma \in \mathcal{F}(M)$ we define a function d on $M \times M$ by $d(x,y) = 0$ if $x - y \in \bigcap \Gamma_v$ while $d(x,y) = 2^{-k}$ when $x - y \in \Gamma_k \setminus \Gamma_{k-1}$.

It is obvious that d is a pseudo-metric on M and it is used to get a topology on M. This topology depends *a priori* on the chosen good filtration. However, using 2.3 below, we see that this topology does not depend on the particular choice of Γ in the class $\mathcal{F}(M)$, so we find an intrinsic topology on the finitely generated R-module M which we call the $\mathcal{F}$-topology.

2.3. LEMMA. *If Γ and $\Omega \in \mathcal{F}(M)$ then there exists an integer w such that $\Gamma_{v-w} \subset \Omega_v \subset \Gamma_{v+w}$ hold for all v.*

Proof. We have $\Omega_v = \Sigma_{v-m_1} u_1 + \cdots + \Sigma_{v-m_s} u_s$ and since any good filtration is exhaustive we find w' so that $u_j \in \Gamma_{w'}$ for all j. If we take $w = \max\{w' - m_1, \ldots, w' - m_s\}$, then $\Omega_v \subset \Gamma_{v+w}$ for all v and so on.

2.4. *Example.* The $\mathcal{F}$-topology on M is Hausdorff if and only if good filtrations on M are separated.

From now on we assume that the ring R is both left and right Noetherian and begin to study the abelian category of finitely generated left (or right) R-modules.

Let M be a finitely generated R-module and let $K \subset M$ be some submodule. If $\Gamma \in \mathcal{F}(M)$ and if $\Omega \in \mathcal{F}(K)$, then there exists some integer w such that $\Omega_v \subset \Gamma_{v+w}$ hold for all v. The proof is easy, i.e., use the method from Lemma 2.3 above.

Conversely, we may ask if there exists some w so that the inclusions $\Gamma_{v-w} \cap K \subset \Omega_v$ hold for all v. It is not true in general so we can impose it as a condition. This leads to

2.5. *Definition.* The filtration on the ring R satisfies the comparison condition if the following is true. For any finitely generated R-module M and any submodule K and any $\Gamma \in \mathcal{F}(M)$ and any $\Omega \in \mathcal{F}(K)$, it follows that there exists some integer w with $\Gamma_{v-w} \cap K \subset \Omega_v$ for all v.

Remark. Here M can be either a left or a right R-module, i.e., the comparison condition is left and right symmetric.

2.6. *A criterion for c.c.* Let R be a filtered ring. Then the filtration satisfies the comparison condition if and only if the integer w exists for pairs $K \subset_R R$ or $K \subset R_R$, i.e., it suffices to take M as the left (or the right) R-module R.

Proof. Assume that w integers exist for pairs $K \subset_R R$, i.e., here K are left ideals in the ring R. It follows easily that integers w exist for pairs $K \subset F$ when F are finitely generated and free R-modules. Consider now a pair $K \subset M$ where M is finitely generated. Then we can find a commutative diagram

$$\begin{array}{ccccc} F' & \xrightarrow{\varphi} & K & \rightarrow & 0 \\ \alpha \downarrow & & \downarrow & & \\ F & \xrightarrow{\psi} & M & \rightarrow & 0 \end{array}$$

where F' and F are free R-modules of finite rank. Observe now that we can find $\Omega \in \mathcal{F}(K)$ and $\Gamma \in \mathcal{F}(M)$ which are images of good filtrations on F' and on F respectively, i.e., $\Omega_v = \varphi(\Gamma_v(F'))$ and $\Gamma_v = \psi(\Gamma_v(F))$ and then we find w so that $\alpha(F') \cap \Gamma_{v-w}(F) \subset \alpha(\Gamma_v(F'))$ and obtain the inclusions $\Gamma_{v-w} \cap K \subset \Omega_v$ for all v.

2.7. *Example.* Let us consider a left ideal L in the ring R. It is a finitely generated R-module since the ring is left Noetherian. If $L = Ru_1 + \cdots + Ru_s$ we find that $\{\Gamma_v = \Sigma_v u_1 + \cdots + \Sigma_v u_s\} \in \mathcal{F}(L)$ and if the comparison condition holds, then $\exists w$ so that the inclusions $\Sigma_{v-w} \cap (Ru_1 + \cdots + Ru_s) \subset \Sigma_v u_1 + \cdots + \Sigma_v u_s$ hold for all v. In the same way we discover inclusions $\Sigma_{v-w} \cap (u_1 R + \cdots + u_s R) \subset u_1 \Sigma_v + \cdots + u_s \Sigma_v$ and hence the comparison condition holds for the filtered ring R if and only if the following is true.

2.8. *A c.c. criterion.* The comparison condition holds if and only if the following is true. For any finite subset $u_1, \ldots, u_s$ in the ring R there exists some w so that $\Sigma_{v-w} \cap (Ru_1 + \cdots + Ru_s) \subset \Sigma_v u_1 + \cdots + \Sigma_v u_s$ and $\Sigma_{v-w} \cap (u_1 R + \cdots + u_s R) \subset u_1 \Sigma_v + \cdots + u_s \Sigma_v$ for all v.

2.9. *The case when* $\operatorname{gr}(R)$ *is Noetherian.* Let R be a filtered Noetherian ring. Put $\operatorname{gr}(R) = \oplus \Sigma_v / \Sigma_{v-1}$, which is called the associated graded ring. We do not assume that $\mathrm{gr}(R)$ is commutative. However, we can impose the condition that $\mathrm{gr}(R)$ also is left and right Noetherian and then one has

2.10. PROPOSITION. *Assume that both R and $\mathrm{gr}(R)$ are Noetherian and that the comparison condition holds. Then good filtrations induce good filtrations on submodules.*

Proof. Let M be a finitely generated R-module and let $K \subset M$ be some submodule. The assertion is that if $\Gamma \in \mathcal{F}(M)$, then $\{\Gamma_v \cap K\} \in \mathcal{F}(K)$. To prove it we put $\Omega_v = \Gamma_v \cap K$ and then $\mathrm{gr}(K) = \oplus\Omega_v/\Omega_{v-1}$ is a $\mathrm{gr}(R)$-submodule of $\mathrm{gr}(M) = \oplus\Gamma_v/\Gamma_{v-1}$. Here $\mathrm{gr}(M)$ is a finitely generated $\mathrm{gr}(R)$-module since Γ is good. It follows that the $\mathrm{gr}(R)$-submodule $\mathrm{gr}(K)$ is finitely generated too. Choose a finite set $u_1, \cdots, u_s$ in K whose symbols generate the $\mathrm{gr}(R)$-module $\mathrm{gr}(K)$. This means that there are integers $m_1, \ldots, m_s$ with $u_j \in \Omega_{m_j}$ and $\mathrm{gr}(K) = \mathrm{gr}(R)\gamma(u_1) + \cdots + \mathrm{gr}(R)\gamma(u_s)$ where $\gamma(u_j)$ denote the images of u_j in $\Omega_{m_j}/\Omega_{m_j-1}$.

An easy recursion gives

SUBLEMMA 1. *$\Omega_v \subset \Sigma_{v-m_1}u_1 + \cdots + \Sigma_{v-m_s}u_s + \Omega_{v-j}$ hold for all $j \geq 1$ and all v.*

At this stage we use the comparison condition applied to the finite subset $u_1, \ldots, u_s$ of M to get some large integer w such that $\Omega_{v-w} \subset \Sigma_{v-m_1}u_1 + \cdots + \sum_{v-m_s} u_s$ for all v and then we find that $\Gamma_v \cap K = \Omega_v = \Sigma_{v-m_1}u_1 + \cdots + \Sigma_{v-m_s}u_s$ belongs to the family $\mathcal{F}(K)$.

2.11. *The closure condition.* Let R be a filtered ring. On the left R-module ${}_RR$ we get the $\mathcal{F}$-topology using the metric $d(x,y) = 2^{-k}$ if $x - y \in \Sigma_k - \Sigma_{k-1}$. An additive subgroup H of R is $\mathcal{F}$-closed when $H = \bigcap_{-\infty}^{+\infty}(H + \Sigma_v)$ and this leads to

2.12. *Definition.* The filtration Σ satisfies the closure condition if the additive subgroups $\Sigma_{v-m_1}u_1 + \cdots + \Sigma_{v-m_s}u_s$ and $u_1\Sigma_{v-m_1} + \cdots + u_s\Sigma_{v-m_s}$ are $\mathcal{F}$-closed for any finite subset $u_1, \ldots, u_s$ in R and all integers $v, m_1, \ldots, m_s$.

Now we can prove

2.13. PROPOSITION. *If Σ satisfies the closure condition and if $\mathrm{gr}(R)$ is a left and a right Noetherian ring then R is left and right Noetherian and Σ satisfies the comparison condition.*

Proof. Let L be a left ideal in the ring R. Let $\sigma(L)$ be the left ideal of $\mathrm{gr}(R)$ which is generated by principal symbols of elements in L. Since $\mathrm{gr}(R)$ is left Noetherian we find a finite subset $u_1, \cdots, u_s$ in L so that $\sigma(L) = \mathrm{gr}(R)\sigma(u_1) + \cdots + \mathrm{gr}(R)\sigma(u_s)$. Let $m_j = \mathrm{ord}(u_j)$ and then a trivial recursion gives the inclusions

$$\Sigma_v \cap L \subset \Sigma_{v-m_1}u_1 + \cdots + \Sigma_{v-m_s}u_s + \Sigma_{v-m} \cap L$$

for each fixed v and all $m \geq 1$. Now we use that $\Sigma_{v-m_1}u_1 + \cdots + \Sigma_{v-m_s}u_s$ are $\mathcal{F}$-closed to get $\Sigma_v \cap L = \Sigma_{v-m_1}u_1 + \cdots + \Sigma_{v-m_s}u_s$ for all v. It follows that

$L = Ru_1 + \cdots + Ru_s$, which proves that the ring R is left Noetherian since L was an arbitrary left ideal. We have also proved that $\{L \cap \Sigma_v\} \in \mathcal{F}(L)$, which means that the (left) comparison condition holds.

2.14. *Remark.* Assume as above that $\mathrm{gr}(R)$ is Noetherian and that Σ satisfies the closure condition. Then it is easy to prove that if M is a finitely generated R-module then its $\mathcal{F}$-topology is Hausdorff, i.e., $\bigcap \Gamma_v = \{0\}$ for any good filtration on M. We leave the detailed verification to the reader.

Summing up, under the hypothesis that $\mathrm{gr}(R)$ is Noetherian we see that the closure condition implies that R is Noetherian and at the same time the comparison condition holds and good filtrations on finitely generated R-modules are separated.

3. Graded free resolutions. Let R be a filtered ring. For the moment we only assume that R and $\mathrm{gr}(R)$ are both Noetherian rings. Let M be a finitely generated left R-module and choose some $\Gamma \in \mathcal{F}(M)$. Then we can construct an R-free resolution of M with special properties.

3.1. PROPOSITION. *We can construct a free resolution* $\cdots \to F_2 \to F_1 \to F_0 \to M \to 0$ *where each* F_j *is a free* R*-module of finite rank. In addition, each* F_j *is equipped with some good filtration* $\{\Gamma_v(F_j)\}$ *and the* R*-linear mappings preserve the filtrations on the modules above, i.e.,* $\Gamma_v(F_j)$ *is sent into* $\Gamma_v(F_{j-1})$ *for all* $j \geq 1$ *and* $\Gamma_v(F_0) \to \Gamma_v(M)$*. In addition, the associated graded complex*

$$\cdots \to \mathrm{gr}(F_1) \to \mathrm{gr}(F_0) \to \mathrm{gr}_\Gamma(M) \to 0$$

is exact and here $\mathrm{gr}(F_j)$ *are free* $\mathrm{gr}(R)$*-modules of finite rank.*

Remark. When the conditions above are satisfied we call $F_.$ a graded free resolution of the given filtered R-module M.

3.2. *How to construct* $F_.$. First $\Gamma_v = \Sigma_{v-v_1}u_1 + \cdots + \Sigma_{v-v_s}u_s$ where $M = Ru_1 + \cdots + Ru_s$ and $v_1, \ldots, v_s$ are some integers. We define $F_0 = R\varepsilon_1 \oplus \cdots \oplus R\varepsilon_s$ and use the good filtration $\Gamma_v(F_0) = \oplus \Sigma_{v-v_j}\varepsilon_j$ and get an exact sequence $0 \to K \to F_0 \to M \to 0$. Here K is a finitely generated R-module and using the induced filtration $\{\Gamma_v \cap K\}$ we find that $0 \to \mathrm{gr}(K) \to \mathrm{gr}(F_0) \to \mathrm{gr}(M) \to 0$ is exact. So both K and $\mathrm{gr}(K)$ are finitely generated modules. We can then choose $w_1, \cdots, w_t$ in K so that $K = Rw_1 + \cdots + Rw_t$ and $\mathrm{gr}(K) = \mathrm{gr}(R)\gamma(w_1) + \cdots + \mathrm{gr}(R)\gamma(w_t)$. Here $\gamma(w_j) \in \Gamma_{v_j}(F_0)/\Gamma_{v_j-1}(F_0)$ for some integers $v_1, \ldots, v_t$. Now we take $F_1 = R\varepsilon_1 \oplus \cdots \oplus R\varepsilon_t$ and use $\Gamma_v(F_1) = \Sigma_{v-v_1}\varepsilon_1 \oplus \cdots \oplus \Sigma_{v-v_t}\varepsilon_t$. Then the reader may repeat the construction and discover why the graded free resolution exists.

3.3. *A warning.* In Proposition 3.1, we find that $F_.$ is an R-free resolution of M and at the same time $\mathrm{gr}(F_.)$ is a $\mathrm{gr}(R)$-free resolution of $\mathrm{gr}_\Gamma(M)$. However, if v is a given integer it is not always true that the complex

$$\cdots \to \Gamma_v(F_1) \to \Gamma_v(F_0) \to \Gamma_v(M) \to 0$$

is exact. In fact, the construction of F_1 already reveals why this can fail. For here $0 \to K \to F_0 \to M \to 0$ was found and then $F_1 \to K \to 0$ was constructed and the reader may discover that

$$\cdots \to \Gamma_v(F_1) \to \Gamma_v(F_0) \to \Gamma_v(M) \to 0$$

are right exact for all v if and only if $\{\Gamma_v(F_0) \cap K\} \in \mathcal{F}(K)$.

3.4. *Remark.* If we know that a good filtration on a f.g. R-module induces good filtrations on its submodules then we can construct a graded free resolution $F_.$ of M where the complexes $\cdots \to \Gamma_v(F_1) \to \Gamma_v(F_0) \to \Gamma_v(M) \to 0$ are exact for all v. This occurs for example when the given filtration satisfies the comparison condition.

3.5. *The filtered complex* $\mathrm{Hom}(F_., N)$. Let M be a f.g. left R-module and choose some $\Gamma \in \mathcal{F}(M)$ and construct some graded free resolution $F_.$ as above. Let N be another f.g. R-module and choose some $\{\Gamma_v(N)\} \in \mathcal{F}(N)$. Now we consider the Hom-complex

$$0 \to \mathrm{Hom}(F_0, N) \to \mathrm{Hom}(F_1, N) \to \cdots$$

where the subscript R is deleted in Hom, i.e., any Hom consists of R-linear mappings. The complex above is filtered when we use

3.6. *Formula.* $\Gamma_k(\mathrm{Hom}(F_j, N)) = \{\varphi \in \mathrm{Hom}(F_j, N) \colon \varphi(\Gamma_v(F_j)) \subset \Gamma_{v+k}(N)$ for all $v\}$.

The differentials in $\mathrm{Hom}(F_., N)$ preserve the filtrations above. So it means that $\mathrm{Hom}(F_., N)$ is a filtered complex and hence we can construct a spectral sequence whose initial term is the associated graded complex $\mathrm{gr}(\mathrm{Hom}(F_., N))$.

Remark. Recall that a spectral sequence can be constructed from any filtered complex, i.e., by the famous construction due to Leray and Koszul. See, for example, [13] or [5: p. 48] for details.

Let $\{E_r^{\cdot} \colon r = 0, 1, 2, \ldots\}$ be the spectral sequence of the filtered complex $\mathrm{Hom}(F_., N)$. So here $0 \to E_r^0 \to E_r^1 \to \cdots$ are complexes for each r. The differentials in $E_r^{\cdot}$ are homogeneous of degree $-r$, i.e., they send $E_r^k(v) \to E_r^{k+1}(v-r)$ for all k and v.

We have $E_0^k = \mathrm{gr}(\mathrm{Hom}(F_k, N))$ and since F_k is a graded free R-module of finite rank, one has

3.7. *Formula.* $E_0^k \cong \mathrm{Hom}_{\mathrm{gr}(R)}(\mathrm{gr}(F_k), \mathrm{gr}(N))$ hold for all k.

The verification is very easy. Now $E_1^k = \mathcal{H}^k(E_0^{\cdot})$ and since $\mathrm{gr}(F_.)$ is a free resolution of the $\mathrm{gr}(R)$-module $\mathrm{gr}(M)$ we get

3.8. *Formula.* $E_1^k \cong \mathrm{Ext}^k_{\mathrm{gr}(R)}(\mathrm{gr}(M), \mathrm{gr}(N))$.

The spectral sequence $\{E_r^{\cdot} \colon r \geq 0\}$ has been found and it remains to analyze when it is convergent. We say that it converges if the following is true: to each integer k there exists some $r = r(k)$ such that $E_r^k \cong E_\infty^k$ hold for all $r > r(k)$. Here $E_\infty^k = \mathrm{gr}(\mathcal{H}^k(\mathrm{Hom}(F_., N)))$ are the associated graded cohomology groups of the filtered complex $\mathrm{Hom}(F_., N)$. Observe that $\mathcal{H}^k(\mathrm{Hom}(F_., N)) = \mathrm{Ext}_R^k(M, N)$ hold. So $E_\infty^k = \mathrm{gr}(\mathrm{Ext}_R^k(M, N))$.

Now we can announce

3.9. PROPOSITION. *If the filtration on R satisfies the comparison condition then the spectral sequence above is convergent.*

Remark. So the conclusion is valid for any pair of finitely generated R-modules M and N where we equip both with good filtrations and choose some graded free resolution of M to get the filtered complex $\mathrm{Hom}_R(F_., N)$.

Proof. Put $Z^j_\infty = \mathrm{Ker}(\mathrm{Hom}(F_j, N) \to \mathrm{Hom}(F_{j+1}, N))$ and let B^j_∞ be the image of the mapping $\mathrm{Hom}(F_{j-1}, N) \to \mathrm{Hom}(F_j, N)$. The convergence of the spectral sequence follows if we, for each fixed k, find some integer $w = w(k)$ such that the following is true. First we have the equalities

$$\Gamma_v(Z^k_\infty) + \Gamma_{v-1}(\mathrm{Hom}(F_k, N)) = \Gamma_v(Z^k_w) + \Gamma_{v-1}(\mathrm{Hom}(F_k, N))$$

where $\Gamma_v(Z^k_\infty) = \Gamma_v(\mathrm{Hom}(F_k, N)) \cap Z^k_\infty$ and $Z^k_w = \{\varphi \in \Gamma_v(\mathrm{Hom}(F_k, N)) :$ its image $d(\varphi)$ in $\mathrm{Hom}(F_{k+1}, N)$ belongs to $\Gamma_{v-w}(\mathrm{Hom}(F_{k+1}, N))\}$.

Of course, we require these equalities for all v. In addition, we have

$$\Gamma_v(B^k_\infty) + \Gamma_{v-1}(\mathrm{Hom}(F_k, N)) = \Gamma_{v-1}(\mathrm{Hom}(F_k, N)) + \Gamma_v(B^k_w)$$

where $\Gamma_v(B^k_\infty) = B^k_\infty \cap \Gamma_v(\mathrm{Hom}(F_k, N))$ and $\Gamma_v(B^k_w) = \{\varphi \in \Gamma_v(\mathrm{Hom}(F_k, N)) :$ φ is the image of some $\psi \in \Gamma_{v+w-1}(\mathrm{Hom}(F_{k-1}, N))\}$.

See, for example, [4: Proposition 4.4] for the easy proof that these equalities imply the convergence of the spectral sequence. Finally, it is easy to discover that the comparion condition gives the required equalities above, i.e., we find sufficiently large integers w for each k.

3.10. *Remark.* In the special case when $N = {}_RR$ and the left R-module R is equipped with the Σ-filtration, then $\mathrm{Hom}(F_., R)$ has an additional structure using the obvious bimodule structure on the ring R. To be precise, the differentials of the complex $\mathrm{Hom}(F_., R)$ are right R-linear where $\mathrm{Hom}(F_k, R)$ become right R-modules using the rule: to any left R-linear mapping $\varphi: F_k \to R$ and any $\alpha \in R$ we let $\varphi\alpha$ be the left R-linear mapping defined by $(\varphi\alpha)(\xi) = \varphi(\xi)\alpha$ for any $\xi \in F_k$.

Then the comparison condition implies that the induced filtrations on the cohomology groups $\mathcal{H}^k(\mathrm{Hom}(F_., R)) \cong \mathrm{Ext}^k_R(M, R)$ are good and, using Lemma 2.3, the reader may discover how the convergence of the spectral sequence is found when $N = {}_RR$. Namely, the whole point is that if $B^k = \mathrm{Im}(\mathrm{Hom}(F_{k-1}, R))$ then the right R-module B^k_∞ is equipped with two good filtrations. One is found using $\mathrm{Hom}(F_{k-1}, R) \to B^k \to 0$ and the other is found using $\Gamma_v(B^k_\infty) = B^k_\infty \cap \Gamma_v(\mathrm{Hom}(F_k, R))$.

3.11. *Some consequences.* If we assume that both R and $\mathrm{gr}(R)$ are Noetherian and that Σ satisfies the comparison condition, then we have seen that the spectral sequence of $\mathrm{Hom}(F_., N)$ converges, where we have chosen some good filtration on N and some graded free resolution of the Γ-filtered R-module M where $\Gamma \in \mathcal{F}(M)$. The convergence implies that the graded cohomology groups $\mathrm{gr}(\mathrm{Ext}^k_R(M, N)) = E^k_\infty \cong E^k_r$ for some large r are isomorphic to subquotients of $E^k_1 = \mathrm{Ext}^k_{\mathrm{gr}(R)}(\mathrm{gr}(M), \mathrm{gr}(N))$.

This will be used in §4. For the moment we observe the following

3.12. COROLLARY. *Assume that both R and* $\operatorname{gr}(R)$ *are Noetherian and that Σ satisfies the closure condition. If $\Gamma \in \mathcal{F}(M)$ and if $\Omega \in \mathcal{F}(N)$ are two good filtrations on a pair of finitely generated R-modules then one has: if* $\operatorname{Ext}^k_{\operatorname{gr}(R)}(\operatorname{gr}_\Gamma(M), \operatorname{gr}_\Omega(N)) = 0$ *for some given integer k then it follows that* $\operatorname{Ext}^k_R(M, N) = 0$.

Proof. First, the hypothesis implies that Σ satisfies the comparison condition and we conclude that if k is given then $\operatorname{Ext}^k_R(M, N)$ is equipped with a filtration, arising from some chosen graded free resolution of the Γ-filtered R-module M. Here $\operatorname{gr}(\operatorname{Ext}^k_R(M, N))$ is a subquotient of $\operatorname{Ext}^k_{\operatorname{gr}(R)}(\operatorname{gr}_\Gamma(M), \operatorname{gr}_\Omega(N))$ so if this Ext-group is zero we get $\operatorname{gr}(\operatorname{Ext}^k_R(M, N)) = 0$. Now the closure condition implies that the induced filtration on $\operatorname{Ext}^k_R(M, N)$ is separated.

3.13. *Remark.* The proof above is a bit sketchy since we have not offered the details which show that the induced filtrations on the cohomology groups of the filtered complex $\operatorname{Hom}(F_., N)$ are separated when $F_.$ is a graded free resolution of the Γ-filtered R-module M. In the special case when $N = {}_RR$, this follows from the fact that these induced filtrations are good on the right R-modules $\operatorname{Ext}^k(M, R)$ and then we use the remark in 2.14. The general case when N is a finitely generated R-module and $\Omega \in \mathcal{F}(N)$ can be handled in a similar way.

4. Filtered Gorenstein rings. Let R be a filtered Noetherian ring. We assume that $\operatorname{gr}(R)$ also is left and right Noetherian and we can impose the condition that $\operatorname{gr}(R)$ is a Noetherian Gorenstein ring. Then one has

4.1. THEOREM. *Assume that* $\operatorname{gr}(R)$ *is a Noetherian Gorenstein ring and that Σ satisfies the closure condition. Then R is a Noetherian Gorenstein ring.*

Proof. Here we can use Corollary 3.12 with $N = {}_RR$. If μ is the injective dimension of $\operatorname{gr}(R)$ we first discover that $\operatorname{Ext}^k_R(M, R) = 0$ for all $k > \mu$ and any finitely generated R-module M. This means that R has a finite injective dimension. It remains to verify the Gorenstein condition on R. So let M be a finitely generated R-module and if $k \geq 0$ we study $\operatorname{Ext}^k_R(M, R)$ and first we prove Condition 1.13, which amounts to showing that $j(\operatorname{Ext}^k_R(M, R)) \geq k$, i.e., that $\operatorname{Ext}^v_R(\operatorname{Ext}^k_R(M, R), R) = 0$ for all $v < k$. To get this we choose some $\Gamma \in \mathcal{F}(M)$ and then a graded free resolution of the Γ-filtered R-module M. As explained in §3 we get the filtered complex $\operatorname{Hom}(F_., R)$ and induced good filtrations on the R-modules $\operatorname{Ext}^k_R(M, R)$. Here the associated graded $\operatorname{gr}(R)$-modules $\operatorname{gr}(\operatorname{Ext}^k_R(M, R))$ are subquotients of $\operatorname{Ext}^k_{\operatorname{gr}(R)}(\operatorname{gr}_\Gamma(M), \operatorname{gr}(R))$.

Now $\operatorname{gr}(R)$ is Gorenstein and the j-numbers of finitely generated $\operatorname{gr}(R)$-modules exist. In particular, $j(\operatorname{Ext}^k_{\operatorname{gr}(R)}(\operatorname{gr}_\Gamma(M), \operatorname{gr}(R))) \geq k$ and the inequality remains for its subquotient $\operatorname{gr}(\operatorname{Ext}^k_R(M, R))$. So if we put $N = \operatorname{Ext}^k_R(M, R)$ we have some good filtration on N for which $j(\operatorname{gr}(N)) \geq k$, which means that $\operatorname{Ext}^v_{\operatorname{gr}(R)}(\operatorname{gr}(N), \operatorname{gr}(R)) = 0$ for all $v < k$. We apply Corollary 3.12 to the R-module N and get $\operatorname{Ext}^v_R(N, R) = 0$ for all $v < k$ which gives $j(\operatorname{Ext}^k_R(M, R)) \geq k$.

Hence Condition 1.13 holds and the Gorenstein condition in 1.16 is proved in a similar way, i.e., if we take a submodule N of $\mathrm{Ext}_R^k(M,R)$, we use the constructions, as above, which first gave a good filtration on $\mathrm{Ext}_R^k(M,R)$ induced from the filtered complex $\mathrm{Hom}(F_\cdot, R)$. It induces a good filtration on the submodule N. Then $\mathrm{gr}(N)$ is some $\mathrm{gr}(R)$-submodule of the $\mathrm{gr}(R)$-module $\mathrm{gr}(\mathrm{Ext}_R^k(M,R))$ and since $\mathrm{gr}(R)$ is Gorenstein we know that $j(\mathrm{gr}(\mathrm{Ext}_R^k(M,R))) \geq k \Rightarrow j(\mathrm{gr}(N)) \geq k$ and then Corollary 3.12 is applied to N and gives $j(N) \geq k$.

4.2. *Remark.* So if we assume that $\mathrm{gr}(R)$ is Gorenstein and that Σ satisfies the closure condition, then the proof above has shown that R is Gorenstein. Also, if M is any finitely generated R-module then $j(M) \geq j(\mathrm{gr}_\Gamma(M))$ holds where Γ is any good filtration on M. It turns out that a stronger results holds.

4.3. THEOREM. *We have $j(M) = j(\mathrm{gr}_\Gamma(M))$ under the conditions given above.*

To prove it we have to show that $j(M) \leq j(\mathrm{gr}_\Gamma(M))$ and this inequality is derived below under an assumption which is even weaker than the imposed closure condition. The result is

4.4. THEOREM. *Assume that both R and $\mathrm{gr}(R)$ are Noetherian Gorenstein rings and that Σ satisfies the comparison condition. Then $j(M) \leq j(\mathrm{gr}_\Gamma(M))$ holds for any finitely generated R-module M and any $\Gamma \in \mathcal{F}(M)$.*

Proof. Let $F_\cdot$ be some graded free resolution of M. In the spectral sequence we have $E_1^v = \mathrm{Ext}_{\mathrm{gr}(R)}^v(\mathrm{gr}(M), \mathrm{gr}(R))$. If $k = j(\mathrm{gr}(M))$ this means that $E_1^v = 0$ for all $0 \leq v < k$.

In particular, no coboundaries occur in degree k and we find an exact sequence $0 \to E_2^k \to E_1^k \to S_1 \to 0$ with $S_1 \subset E_1^{k+1}$.

Passing to higher terms we get exact sequences

$$0 \to E_{r+1}^k \to E_r^k \to S_r \to 0 \quad \text{with} \quad S_r \subset E_r^{k+1}$$

for all $r \geq 2$.

Now $E_1^{k+1} = \mathrm{Ext}_{\mathrm{gr}(R)}^{k+1}(\mathrm{gr}(M), \mathrm{gr}(R))$ is a $\mathrm{gr}(R)$-module whose j-number is $\geq k+1$. If $r > 1$ we know that E_r^{k+1} is a subquotient of E_1^{k+1} and hence $j(E_r^{k+1}) \geq k+1$ and similarly $j(S_r) \geq k+1$ above.

If r is sufficiently large we have $E_r^k = E_{r+1}^k = \mathrm{gr}(\mathrm{Ext}_R^k(M,R))$ and if k were $< j(M)$ we would have $E_r^k = 0$. Starting from this we count j-integers in the exact sequences above and by an obvious descending induction we get $j(E_1^k) \geq k+1$. This is a contradiction since $k = j(\mathrm{gr}_\Gamma(M))$ gives $j(E_1^k) = k$ by Proposition 1.18, for example.

5. The case when $\mathrm{gr}(R)$ is commutative. Let R be a filtered ring and assume that $\mathrm{gr}(R)$ is commutative. If M is a finitely generated R-module and if $\Gamma \in \mathcal{F}(M)$ we get the $\mathrm{gr}(R)$-module $\mathrm{gr}(M)$. Put $I_\Gamma(M) = \{\alpha \in \mathrm{gr}(R) \colon \alpha\, \mathrm{gr}_\Gamma(M) = 0\}$. The ideal $I_\Gamma(M)$ depends on the particular good filtration Γ. However, if we take its radical ideal $\sqrt{I_\Gamma(M)}$ then we get an invariantly defined ideal. More precisely we have

5.1. PROPOSITION. *If* Γ *and* $\Omega \in \mathcal{F}(M)$ *then* $\sqrt{I_\Gamma(M)} = \sqrt{I_\Omega(M)}$.

Proof. There exists some w so that $\Omega_{v-w} \subset \Gamma_v \subset \Omega_{v+w}$ hold for all v. Observe that $I_\Gamma(M)$ is a graded ideal in the graded ring $\mathrm{gr}(R)$. It follows that $\sqrt{I_\Gamma(M)}$ also is a graded ideal, which means that it is generated by homogeneous elements in $\mathrm{gr}(R)$. Let $\alpha \in \sqrt{I_\Gamma(M)}$ be homogeneous. Hence $\alpha = \sigma(x)$ for some $x \in R$. If $k = \mathrm{ord}(x)$, i.e., $x \in \Sigma_k - \Sigma_{k-1}$ we observe that $\alpha^s = \sigma(x)^s = \sigma_{ks}(x^s) =$ the image of x^s in $\Sigma_{ks}/\Sigma_{ks-1}$. To say that $\alpha^s \in I_\Gamma(M)$ means that when x^s operates on the filtered R-module M then $x^s\Gamma_v \subset \Gamma_{v+ks-1}$ hold for all v. It follows that $x^{2s}\Gamma_v \subset \Gamma_{v+2ks-2}$ and so on. We get $x^{s(2w+1)}\Gamma_v \subset \Gamma_{(2w+1)sk+v-2w-1}$ for all v. This gives that $x^{s(2w+1)}\Omega_v \subset \Omega_{(2w+1)sk+v-1}$ for all v and it follows that $\alpha^{s(2w+1)}$ belongs to $I_\Omega(M)$. This gives $\sqrt{I_\Gamma(M)} \subset \sqrt{I_\Omega(M)}$ and so on.

5.2. *Definition.* If M is a finitely generated R-module then the unique radical ideal $\sqrt{I_\Gamma(M)}$ found from any $\Gamma \in \mathcal{F}(M)$ is denoted by $J(M)$ and it is called the *characteristic ideal* of M.

From now on we assume that the commutative ring $\mathrm{gr}(R)$ is Noetherian. If M is a finitely generated R-module (left or right) we have found the ideal $J(M)$. This radical ideal has its usual prime decomposition, i.e., $J(M) = \mathcal{P}_1 \cap \cdots \cap \mathcal{P}_s$ where $\mathcal{P}_1, \ldots, \mathcal{P}_s$ are the unique minimal prime divisors.

5.3. *Remark.* If $\Gamma \in \mathcal{F}(M)$ we use that $J(M)$ is the radical of $I_\Gamma(M)$ and conclude that if $\mathrm{gr}_\Gamma(M)_\mathcal{P}$ is the $\mathrm{gr}(R)_\mathcal{P}$-module which arises after a localization at some $\mathcal{P} \in \mathrm{Spec}(\mathrm{gr}(R))$ = the family of prime ideals in $\mathrm{gr}(R)$, then $J(M)_\mathcal{P} = \mathrm{gr}(R)_\mathcal{P} \otimes_{\mathrm{gr}(R)} J(M)$ is the radical of the ideal in $\mathrm{gr}(R)_\mathcal{P}$ which annihilates $\mathrm{gr}_\Gamma(M)_\mathcal{P}$.

Of course, this uses that $\mathrm{gr}(R)_\mathcal{P}$ is a flat $\mathrm{gr}(R)$-module. If $\mathcal{P}$ is a minimal prime divisor of $J(M)$ then $J(M)_\mathcal{P} = \mathcal{P}\,\mathrm{gr}(R)_\mathcal{P}$ = the maximal ideal of the local ring $\mathrm{gr}(R)_\mathcal{P}$. This gives

5.4. *Formula.* If $\Gamma \in \mathcal{F}(M)$ and if $\mathcal{P}$ is a minimal prime divisor of $J(M)$ then the $\mathrm{gr}(R)_\mathcal{P}$-module $\mathrm{gr}_\Gamma(M)_\mathcal{P}$ is annihilated by some power of the maximal ideal $\mathcal{P}\,\mathrm{gr}(R)_\mathcal{P}$. This means that the $\mathrm{gr}(R)_\mathcal{P}$-module $\mathrm{gr}_\Gamma(M)_\mathcal{P}$ is zero-dimensional. In particular, this leads to

5.5. *Definition.* If $\mathcal{P}$ is a minimal prime divisor of $J(M)$ and if $\Gamma \in \mathcal{F}(M)$ then $\mathcal{L}_\Gamma(M, \mathcal{P})$ denotes the length of the $\mathrm{gr}(R)_\mathcal{P}$-module $\mathrm{gr}_\Gamma(M)_\mathcal{P}$.

Now one can prove that $\mathcal{L}_\Gamma(M, \mathcal{P})$ only depends on M and $\mathcal{P}$, i.e., that $\mathcal{L}_\Gamma(M, \mathcal{P}) = \mathcal{L}_\Omega(M, \mathcal{P})$ for any pair Γ and Ω in $\mathcal{F}(M)$. The proof is not too difficult. It uses a similar technique as in §6.4 below.

For example, if Γ and $\Omega \in \mathcal{F}(M)$ satisfy $\Gamma_v \subset \Omega_v \subset \Gamma_{v+1}$ for all v, then the equality $\mathcal{L}_\Gamma(M, \mathcal{P}) = \mathcal{L}_\Omega(M, \mathcal{P})$ for any minimal prime divisor $\mathcal{P}$ of $J(M)$ is easily found and the general case is handled as in §6.4.

Summing up, in addition to $J(M)$ we get more invariants by

5.6. *Definition.* To any minimal prime divisor $\mathcal{P}$ of $J(M)$ we get the unique integer $e(M, \mathcal{P}) = \mathcal{L}_\Gamma(M, \mathcal{P})$ for any $\Gamma \in \mathcal{F}(M)$. We call the positive integer $e(M, \mathcal{P})$ the multiplicity of M along $\mathcal{P}$.

5.7. *Remark.* Using the multiplicities we get a more refined invariant than $J(M)$ using the "cycle" $\sum e(M, \mathcal{P})$ with $\sum$ taken over the minimal prime divisors of $J(M)$. In addition we have a "counting with multiplicities." Namely, define $e(M, \mathcal{P})$ for any $\mathcal{P} \in \mathrm{Spec}(R)$ where the rule is that $e(M, \mathcal{P}) = 0$ if $\mathcal{P}$ is not a minimal prime divisor of $J(M)$ and otherwise it is the multiplicity from Definition 5.6. Then one has

5.8. *Formula.* If $0 \to N \to M \to K \to 0$ is an exact sequence of finitely generated R-modules then $e(M, \mathcal{P}) = e(N, \mathcal{P}) + e(K, \mathcal{P})$ for any prime $\mathcal{P}$.

So far we have only assumed that $\mathrm{gr}(R)$ is a commutative Noetherian ring. Let us now assume more, i.e., that both R and $\mathrm{gr}(R)$ are Noetherian Gorenstein rings. If M is a finitely generated R-module we find the R-modules $\mathrm{Ext}^k_R(M, R)$ for all $k \geq 0$ and try to relate $J(M)$ with characteristic ideals of these R-modules. The following result is quite useful. *Note.* In §5.9 we assume that the filtration Σ on R satisfies the comparison condition.

5.9. PROPOSITION. *Let $\mathcal{P}$ be a minimal prime divisor of $J(M)$ and put $k = \mathrm{inj.dim}(\mathrm{gr}(R)_\mathcal{P})$. Then $\mathcal{P}$ is a minimal divisor of $J(\mathrm{Ext}^k_R(M, R))$.*

Proof. Choose some $\Gamma \in \mathcal{F}(M)$ and a graded free resolution $F_{\cdot}$ of the Γ-filtered R-module M. Consider the complex $\mathrm{Hom}_{\mathrm{gr}(R)}(\mathrm{gr}(F_{\cdot}), \mathrm{gr}(R)) = E_0^{\cdot}$ in the spectral sequence attached to the filtered complex $\mathrm{Hom}(F_{\cdot}, R)$.

We can localize the $\mathrm{gr}(R)$-complex $E_0^{\cdot}$ with respect to $\mathcal{P}$ and get the complex $0 \to (E_0^0)_\mathcal{P} \to (E_0^1)_\mathcal{P} \to \cdots$ where E_0^k are $\mathrm{gr}(R)$-modules which later are localized with respect to the prime $\mathcal{P}$.

Since $\mathrm{gr}(R)_\mathcal{P}$ is a flat $\mathrm{gr}(R)$-module, we see that the cohomology groups of the complex $(E_0^{\cdot})_\mathcal{P}$ equal the localizations at $\mathcal{P}$ of the cohomology groups $E_1^v = \mathrm{Ext}^v_{\mathrm{gr}(R)}(\mathrm{gr}_\Gamma(M), \mathrm{gr}(R))$.

Hence $\mathcal{H}^v((E_0^{\cdot})_\mathcal{P}) = (E_1^v)_\mathcal{P}$ and since $\mathrm{gr}_\Gamma(M)$ is a finitely generated $\mathrm{gr}(R)$-module, we find that the localizations of its Ext-groups are equal to Ext-groups which are found in the $\mathrm{gr}(R)_\mathcal{P}$-module category. More precisely we have

Formula. $(E_1^v)_\mathcal{P} \cong \mathrm{Ext}^v_{\mathrm{gr}(R)_\mathcal{P}}(\mathrm{gr}_\Gamma(M)_\mathcal{P}, \mathrm{gr}(R)_\mathcal{P})$ hold for all v.

Now we use that $\mathcal{P}$ is a minimal prime divisor of $J(M)$, which implies that the $\mathrm{gr}(R)_\mathcal{P}$-module $\mathrm{gr}_\Gamma(M)_\mathcal{P}$ is zero-dimensional. With k equal to the injective dimension of the Gorenstein ring $\mathrm{gr}(R)_\mathcal{P}$ and recalling here that localizations of commutative Gorenstein rings again are Gorenstein, we get

SUBLEMMA 1. $(E_1^v)_\mathcal{P} = 0$ *for all* $v \neq k$.

Hence the spectral sequence $\{(E_r^{\cdot})_\mathcal{P} : r = 0, 1, 2, \ldots\}$, which is obtained from $\{E_r^{\cdot} : r = 0, 1, 2, \ldots\}$ after localization, degenerates. We get $(E_1^k)_\mathcal{P} \cong (E_r^k)_\mathcal{P}$ for all $r \geq 2$ and also $(E_r^v)_\mathcal{P} = 0$ for all $v \neq k$.

Passing to the limit we have $E_r^k = \mathrm{gr}(\mathrm{Ext}^k(M, R))$ as before and we perform a localization at $\mathcal{P}$ and use $(E_1^k)_\mathcal{P} = (E_r^k)_\mathcal{P}$ for a large r and obtain

SUBLEMMA 2. *$(\mathrm{gr}(\mathrm{Ext}^v(M,R)))_{\mathcal{P}} = 0$ for all $v \neq k$ while the $\mathrm{gr}(R)_{\mathcal{P}}$-modules $(\mathrm{gr}(\mathrm{Ext}^k(M,R)))_{\mathcal{P}}$ and $\mathrm{Ext}^k_{\mathrm{gr}(R)_{\mathcal{P}}}(\mathrm{gr}_\Gamma(M)_{\mathcal{P}}, \mathrm{gr}(R)_{\mathcal{P}})$ are isomorphic.*

This means, in particular, that if we use the good filtration on the R-module $\mathrm{Ext}^k(M,R)$ arising from the chosen graded free resolution of the Γ-filtered module M, then the $\mathrm{gr}(R)_{\mathcal{P}}$-module $(\mathrm{gr}(\mathrm{Ext}^k(M,R)))_{\mathcal{P}}$ is zero-dimensional and not identically zero. This means precisely that $\mathcal{P}$ is a minimal prime divisor of $J(\mathrm{Ext}^k(M,R))$.

5.10. *Remark.* Observe that the proof above also gives that the multiplicities $e(M,\mathcal{P})$ and $e(\mathrm{Ext}^k(M,R),\mathcal{P})$ are equal for any minimal prime divisor $\mathcal{P}$ of $J(M)$ with $k = \mathrm{inj.dim}(\mathrm{gr}(R)_{\mathcal{P}})$.

Using Proposition 5.9 we can draw several consequences. For example, we get

5.11. COROLLARY. *Let $k = \mathrm{inj.dim}(\mathrm{gr}(R)_{\mathcal{P}})$ for some minimal prime divisor $\mathcal{P}$ of $J(M)$. Then $\mathrm{Ext}^k_R(\mathrm{Ext}^k_R(M,R),R) \neq 0$.*

Proof. Here $\mathrm{gr}(R)_{\mathcal{P}}$ is a k-dimensional local Gorenstein ring. This implies that $\mathrm{Ext}^k_{\mathrm{gr}(R)_{\mathcal{P}}}(N, \mathrm{gr}(R)_{\mathcal{P}}) \neq 0$ for any nonvanishing zero-dimensional $\mathrm{gr}(R)_{\mathcal{P}}$-module. During the proof of Proposition 5.9 we have found that $\mathrm{gr}(\mathrm{Ext}^k_R(M,R))_{\mathcal{P}}$ is a nonvanishing and zero-dimensional $\mathrm{gr}(R)_{\mathcal{P}}$-module. Now a localization at $\mathcal{P}$ commutes with the Hom-functor on finitely generated $\mathrm{gr}(R)$-modules so we conclude that $(\mathrm{Ext}^k_{\mathrm{gr}(R)}(\mathrm{gr}(\mathrm{Ext}^k_R(M,R))), \mathrm{gr}(R))_{\mathcal{P}} \neq 0$ and *a fortiori* the $\mathrm{gr}(R)$-module $\mathrm{Ext}^k_{\mathrm{gr}(R)}(\mathrm{gr}(\mathrm{Ext}^k_R(M,R)), \mathrm{gr}(R)) \neq 0$.

This means that the j-number attached to the $\mathrm{gr}(R)$-module $\mathrm{gr}(\mathrm{Ext}^k_R(M,R))$ is $\leq k$. On the other hand, we know that the $\mathrm{gr}(R)$-module $\mathrm{gr}(\mathrm{Ext}^k_R(M,R))$ is a subquotient of $\mathrm{Ext}^k_{\mathrm{gr}(R)}(\mathrm{gr}(M), \mathrm{gr}(R))$ and the last $\mathrm{gr}(R)$-module has j-number $\geq k$ since $\mathrm{gr}(R)$ is Gorenstein. The same inequality holds for its subquotient and we conclude that $j(\mathrm{gr}(\mathrm{Ext}^k_R(M,R))) = k$. Using Theorem 4.4 it follows that the R-module $\mathrm{Ext}^k_R(M,R)$ has j-number $\leq k$. At the same time $j(\mathrm{Ext}^k_R(M,R)) \geq k$ since R is Gorenstein. Hence $j(\mathrm{Ext}^k_R(M,R)) = k$ which gives Corollary 5.11.

5.12. *A special case.* Let M be a finitely generated R-module and assume that it has a pure δ-dimension k. Then $J(M)$ is equi-dimensional, i.e., if $\mathcal{P}$ is any minimal prime divisor of $J(M)$ then the equality $j(M) = \mathrm{inj.dim}(\mathrm{gr}(R)_{\mathcal{P}})$ holds.

In fact, here we apply Corollary 5.11 and Corollary 1.22. combined with the actual proof of Corollary 5.11.

6. A study of $J(M/\varphi(M))$.

In this section we assume that R is a filtered ring which is Noetherian and Gorenstein and that $\mathrm{gr}(R)$ is a commutative Noetherian Gorenstein ring. As usual we assume that the filtration satisfies the comparison condition.

Let M be a finitely generated R-module. If $\varphi \in \mathrm{Hom}_R(M,M)$ we find the R-module $M/\varphi(M)$ and our aim is to prove

6.1. THEOREM. *Assume that* φ *is injective and that* $M/\varphi(M)$ *is not the zero module and that* M *has a pure* δ*-dimension. Then* $J(M/\varphi(M))$ *is equidimensional and* $\text{inj.dim}(\text{gr}(R)_{\mathcal{P}}) = j(M) + 1$ *for any minimal prime divisor of* $J(M/\varphi(M))$. *Also, the* R*-module* $M/\varphi(M)$ *has* $j(M/\varphi(M)) = j(M) + 1$.

The proof requires several steps. First, we need some preliminary constructions which serve as tools during the proof. Namely, let T be a new variable so that $R[T]$ is the polynomial ring with coefficients in R. Now M becomes a left $R[T]$-module using the rule $Tu = \varphi(u)$ for any $u \in M$ and, more generally, $P(T)u = P(\varphi)u$ for any $P(T) = p_0 + p_1T + \cdots + p_sT^s \in R[T]$.

Since the left R-linear mapping φ is injective on M we see that the kernel of T is zero, i.e., T is a nonzero divisor on the $R[T]$-module M. Hence the localization $M[T^{-1}] = R[T, T^{-1}] \otimes_R M$ exists. Here $R[T, T^{-1}]$ is the ring of finite Laurent series in the variable T.

Observe that $M \subset T^{-1}M \subset T^{-2}M$ increases inside $M[T^{-1}]$ and it is strictly increasing too. In fact, if $T^{-n}M = T^{-n-1}M$ for some integer n, we multiply by T^{n+1} and obtain $M = TM = \varphi(M)$, which contradicts the assumption that $M \neq \varphi(M)$. So *the* R*-module* $M[T^{-1}]$ is *not* Noetherian.

Observe also that $M/\varphi(M) = M/TM$ and the last R-module is isomorphic to the R-module $T^{-1}M/M$. The proof of Theorem 6.1 relies on the following two results.

6.2. LEMMA. *Let* $\mathcal{C}$ *be the family of finitely generated* R*-modules* M' *inside* $M[T^{-1}]$ *satisfying:* $M \subset M'$ *and* $\delta(M'/M) \leq \delta(M) - 2$. *Then* $\mathcal{C}$ *is a Noetherian family of* R*-modules.*

6.3. LEMMA. *If* M' *is a maximal module in the family* $\mathcal{C}$ *then the* R*-module* $T^{-1}M'/M'$ *has a pure* δ*-dimension which equals* $\delta(M) - 1$.

Remark. Therefore, Lemma 6.2 means that if $M_1 \subset M_2 \subset M_3 \subset \cdots$ for an increasing sequence of R-submodules of $M[T^{-1}]$ which all belong to $\mathcal{C}$ then the sequence is stationary.

In 6.6 below we will show how Lemma 6.2 leads to Theorem 6.1.

Proof of Lemma 6.2. Recall that if $\mu = \text{inj.dim}(R)$ then $\delta(N) + j(N) = \mu$ for any finitely generated R-module N. So if $N \in \mathcal{C}$ the inequality $\delta(N/M) \leq \delta(M) - 2$ means that $j(N/M) \geq j(M) + 2$.

If we look at the long exact sequence of Ext-groups, using R as a second factor, arising from $0 \to M \to N \to M/N \to 0$ with $N \in \mathcal{C}$, then we get $\text{Ext}^k(M, R) \cong \text{Ext}^k(N, R)$, with $k = j(M)$.

Next, $N \subset M[T^{-1}]$ and, since N is a finitely generated R-module, there exists some large n so that $N \subset T^{-n}M$. The two R-modules M and $T^{-n}M$ are isomorphic. Since M has a pure δ-dimension, $T^{-n}M$ has it and hence also its submodule N has it. The pure δ-dimensionality of N implies that N is isomorphic to a submodule of $\text{Ext}^k(\text{Ext}^k(N, R), R)$ where we have used $k = j(M) = j(N)$.

Let us put $M^{\times} = \operatorname{Ext}^k(\operatorname{Ext}^k(M,R),R)$. Since $\operatorname{Ext}^k(M,R) \cong \operatorname{Ext}^k(N,R)$ we get $M^{\times} \cong \operatorname{Ext}^k(\operatorname{Ext}^k(N,R),R)$, which means that there exists an injective R-linear mapping from N into $M^{\times}$.

Suppose now that $M_1 \subset M_2 \subset \cdots$ increase in the family $\mathcal{C}$. To each v we find some $\alpha_v \in \operatorname{Hom}_R(M_v, M^{\times})$ which is injective. If we have found these mappings so that $\alpha_{v+1}|M_v = \alpha_v$ hold for all v, then we see that $\alpha_1(M_1) \subset \alpha_2(M_2) \subset \cdots$ increase inside the Noetherian R-module $M^{\times}$ and hence this sequence is stationary. Since all α_v are injective we find that the sequence $M_1 \subset M_2 \subset \cdots$ is stationary too.

To construct α_v satisfying the compatibility conditions above we first study the special increasing sequence $M \subset T^{-1}M \subset T^{-2}M$ where we have injective mappings $\beta_n: T^{-n}M \to M^{\times}$ for the same reasons as above. They can be constructed to satisfy $\beta_{n+1}|T^{-n}M = \beta_n$ for all n. Then we take some M_v and choose n large to have $M_v \subset T^{-n}M$ and use $\alpha_v = \beta_n|M_v$.

The construction of β_n. Take a projective resolution of M and find a realization of $\mathbb{R}\operatorname{Hom}(\mathbb{R}\operatorname{Hom}(M,R),R)$ as in §1, using some projective resolution $Q_{..}$ of $\hat{P}_{.}$ and then pass to $\hat{Q}_{..}$ and its simple diagonal complex $\Delta^{\cdot}$. If $n > 1$ we find that the multiplication by T^{-n} gives $T^{-n}\Delta^{\cdot} = \mathbb{R}\operatorname{Hom}(\mathbb{R}\operatorname{Hom}(T^{-n}M,R),R)$.

We can simply consider $\Delta^{\cdot}$ as a subcomplex of $T^{-n}\Delta^{\cdot}$ and the second filtration on $T^{-n}\Delta^{\cdot}$ indexes the second filtration on $\Delta^{\cdot}$. Passing to the second terms of the corresponding spectral sequences, where the formula in 1.4 is used, we obtain a commutative diagram

$$\begin{array}{ccc} M & \to & M^{\times} \\ \downarrow & & \downarrow \\ T^{-n}M & \to & \operatorname{Ext}^k(\operatorname{Ext}^k(T^{-n}M,R),R) \end{array}$$

and at this stage the reader discovers that canonical mappings β_n from $T^{-n}M$ into $M^{\times}$ can be found to satisfy $\beta_{n+1}|T^{-n}M = \beta_n$ for all n.

Proof of Lemma 6.3. First we have $0 \to M' \to T^{-1}M' \to T^{-1}M'/M' \to 0$ and here the R-modules M' and $T^{-1}M'$ are isomorphic. A counting with multiplicities implies that $\delta(T^{-1}M'/M') < \delta(M') = \delta(M)$.

It remains to prove that $\delta(S) \geq \delta(M) - 1$ for any nonvanishing R-submodule of $T^{-1}M'/M'$. Now any such S corresponds to a finitely generated R-module N inside $M[T^{-1}]$ with $M' \subset N \subset T^{-1}M'$. Here $M' \neq N$ and the maximality of M' in $\mathcal{C}$ implies $\delta(N/M') > \delta(M) - 2$. Since $\delta(M'/M) \leq \delta(M) - 2$ we conclude that $\delta(N/M') \geq \delta(M) - 1$ and now we use that $S \cong N/M'$.

6.4. *A useful equality.* Keeping the notations as above we consider some finitely generated R-module N inside $M[T^{-1}]$ where $N \supset M$. Then we can prove

6.5. PROPOSITION. $J(M/\varphi(M)) = J(T^{-1}N/N)$.

Proof. Since N is finitely generated there exists some n so that $N \subset T^{-n}M$ and we are going to use an induction over n. If $n = 1$ we have $M \subset N \subset T^{-1}M$

and obtain two exact sequences:

$$0 \to T^{-1}M/N \to T^{-1}N/N \to T^{-1}N/T^{-1}M \to 0,$$

$$0 \to N/M \to T^{-1}M/M \to T^{-1}M/N \to 0.$$

Now we use that $N/M \cong T^{-1}N/T^{-1}M$ and get $J(T^{-1}N/N) = J(T^{-1}M/M) = J(M/\varphi(M))$.

If $M \subset N \subset T^{-2}M$ we consider $N_1 = T^{-1}M + N$. Here $T^{-1}M \subset N_1 \subset T^{-2}M$ is used to get $J(M/\varphi(M)) = J(T^{-1}N_1/N_1)$. We also have $N \subset N_1 \subset T^{-1}N$ and use the same proof as above to get $J(T^{-1}N/N) = J(T^{-1}N_1/N_1)$ and the induction goes on.

6.6. *Proof of Theorem* 6.1. First we use Lemma 6.2 and find a maximal module M' in the family $\mathcal{C}$. Now Lemma 6.3 and §5.12 show that $J(T^{-1}M'/M')$ is equidimensional with $\text{inj.dim}(\text{gr}(R)_{\mathcal{P}}) = j(M) + 1$ for all minimal prime divisors of $J(T^{-1}M'/M')$. Finally, use Proposition 6.5.

II. Gabber's Integrability Theorem

Summary. If R is a filtered ring such that $\text{gr}(R)$ is commutative we can define the *Poisson product* between homogeneous elements in $\text{gr}(R)$. Namely, if x and $y \in R$ are given we find their principal symbols $\sigma(x) \in \Sigma_k/\Sigma_{k-1}$ and $\sigma(y) \in \Sigma_m/\Sigma_{m-1}$ where $k = \text{ord}(x)$ and $m = \text{ord}(y)$. Since $\text{gr}(R)$ is commutative we have $\sigma(xy) = \sigma(x)\sigma(y) = \sigma(y)\sigma(x) = \sigma(yx)$, which in the ring R means that the commutator $xy - yx \in \Sigma_{k+m-1}$ and the Poisson product between $\sigma(x)$ and $\sigma(y)$ is the image of $xy - yx$ in $\Sigma_{k+m-1}/\Sigma_{k+m-2}$. It extends by additivity to give a product on $\text{gr}(R)$ which we denote by $\{\ ,\ \}$ and one verifies that $\text{gr}(R)$ is a Lie ring under $\{\ ,\ \}$.

If L is a left ideal in R and if $\sigma(L)$ is the ideal which is generated by its principal symbols, then it is easily seen that $\{\sigma(L), \sigma(L)\} \subset \sigma(L)$ because if x and $y \in L$ then $xy - yx \in L$. However, if we take the radical $\sqrt{\sigma(L)}$, which is the characteristic ideal of the cyclic module R/L, then the closedness of $\sqrt{\sigma(L)}$ under the Poisson product is not obvious at all.

In this section we prove that $\{J(M), J(M)\} \subset J(M)$ for any finitely generated R-module M under the assumption that $\text{gr}(R)$ is a commutative Noetherian ring and in addition R is a Q-algebra, i.e., it contains the field of rational numbers.

The proof is taken from Gabber's work [10] and we offer a detailed review of [10]. The result above is actually included in a more general case where the filtered rings appear as special cases of so called "T-rings". A "T-ring" is a ring R which contains a distinguished element T such that the one-sided principal ideals RT and TR are equal and the quotient ring $R/(T)$ is commutative. Here $(T) = RT = TR$ is the 2-sided ideal generated by T. In §1 we find the additional conditions which are needed to define a Poisson product on $R/(T)$ and then Theorem 1.3 is announced. It asserts that if the T-ring R is a Q-algebra and if $R/(T)$ is Noetherian then $\{J(M), J(M)\} \subset J(M)$ for any finitely generated

R-module M. Here $J(M)$ is defined as the radical of the ideal in $R/(T)$ which annihilates M/TM.

The proof requires a number of steps. First we reduce it to the special case when $T^2 = 0$. The commutativity of $R/(T)$ and $T^2 = 0$ implies that "3-fold commutators" are zero. This implies that certain *noncommutative localizations* exist and are realized by Ore quotients. So in §2 we are able to reduce the proof of Theorem 1.3 to the special case when $R/(T)$ is a local Noetherian ring and M/TM is zero-dimensional.

Then we find an "Artinian reduction" and the final part of the proof uses some computations of matrices and ends up with a *trace formula*. To be precise, the final step uses that if A and B are two matrices with coefficients in a commutative field then AB and BA have equal traces.

Summing up, we offer a self-contained proof including background about noncommutative localizations. Of course, all the ingredients are already found in Gabber's original work [10] so we have simply offered a review of his ingenious proof.

Final Remarks. For special filtered rings such as the Weyl algebra $A_n(\mathbb{C})$ equipped with its "Σ-filtration" which we define in Part III, the inclusion $\{J(M), J(M)\} \subset J(M)$ was already proved in [28] by other methods. More precisely, the proof in [28] employs micro-local differential operators and some analysis and we also refer to [24] for a proof which still uses methods from [28] but also makes use of a trace formula as we do in §4. Let us also remark that Gabber's result, which really is given in Theorem 1.3, can be used to settle problems which were not known by the more micro-local technique in [28]. For example, Gabber's proof can be used to prove that the irregular part of the support of a coherent $\mathcal{E}$-module is involutive. See [18] for a discussion about this.

1. ***T*-rings.** A ring R is called a T-ring when R contains an element T satisfying the following conditions:

(1) the one-sided principal ideals RT and TR are equal;

(2) $R/(T)$ is a commutative ring where $(T) = RT = TR$ is the 2-sided ideal which T generates by (1);

(3) both $\{x \in R: Tx = 0\}$ and $\{y \in R: yT = 0\}$ are $\subset (T)$;

(4) $Tx - xT \in RT^2$ for all x in R.

1.1. *The Poisson product on $R/(T)$.* If $x \in R$ we denote its image in $R/(T)$ by $\overline{x}$. Let ξ and η be two elements in $R/(T)$. We take x and y in R so that $\xi = \overline{x}$ and $\eta = \overline{y}$. Since $R/(T)$ is commutative we have $xy - yx$ in (T). So in the ring R we can find an element u such that $xy - yx = Tu$ where we have used $(T) = TR$ from (1). The image $\overline{u}$ is called the Poisson product of ξ and η.

Remark. Above we put $\{\xi, \eta\} = \overline{u}$ and it remains only to verify that $\overline{u}$ only depends on ξ and η. First, if $\xi = \overline{x}$ and if $\eta = \overline{y}$ and if we solve $xy - yx = Tu'$ with another element u' in R then we have $T(u - u') = 0$ and hence $u - u' \in (T)$ by (3) so the images of u and u' are equal in $R/(T)$. Next, if $\xi = \overline{z}$ for another element z in the ring R and if $zy - yz = Tw$, we can use (4) to get $\overline{u} = \overline{w}$. In fact, $xy - yx - (zy - yz) = (x - z)y - y(x - z)$ and here $x - z \in (T)$ so that $x - z = \varphi T$ for some $\varphi \in R$.

Now $\varphi Ty - y\varphi T = \varphi(Ty - yT) + (\varphi y - y\varphi)T \in RT^2$ which gives $T(u - w) \in RT^2 = T^2 R$ and using (3) we get $\overline{u} = \overline{w}$ as required.

Summing up, we have seen that the conditions on the element T imply that a product $\{\xi, \eta\}$, which we call the Poisson product, is defined on $R/(T)$. It is easily seen that $\{\ ,\ \}$ satisfies *Jacobi's rule* so $R/(T)$ is a Lie ring under $\{\ ,\ \}$.

If M is a left R-module we get the subset TM and using (1) we see that TM is an R-submodule of M and M/TM is a module over the commutative ring $R/(T)$. This leads to

1.2. *Definition.* $J(M)$ is the radical ideal in $R/(T)$ which equals the radical of the annihilating ideal $I(M/TM) = \{\xi \in R/(T) \colon \xi(M/TM) = 0\}$.

We call $J(M)$ the characteristic ideal of the R-module M and now we can announce Gabber's integrability theorem for T-rings.

1.3. THEOREM. *Let R be a T-ring which, in addition, is a Q-algebra where Q is the field of rational numbers and assume that $R/(T)$ is Noetherian. Then $\{J(M), J(M)\} \subset J(M)$ holds for any left R-module M such that M/TM is a finitely generated $R/(T)$-module and $\{u \in M : Tu = 0\} \subset TM$.*

Remark. To say that R is a Q-algebra simply means that R contains Q as a central subfield.

The proof of Theorem 1.3 requires several steps. It consists of several "reductions" to more special cases.

1.4. *We can assume that* $T^2 = 0$. If R is a T-ring we observe that Condition (1) gives $RT^2 = TRT = T^2R =$ a 2-sided ideal, which we denote by (T^2), and if $A = R/(T^2)$ and if $\mathcal{T}$ is the image of T in $R/(T^2)$ then A is a $\mathcal{T}$-ring and here $A/(\mathcal{T})$ and $R/(T)$ are canonically isomorphic.

Next, if M is a left R-module we get the left A-module $M/T^2M = \mathcal{M}$ and $J(M) = J(\mathcal{M})$ where $J(\mathcal{M})$ is the characteristic ideal of the A-module $\mathcal{M}$. Also, the Poisson product on $A/(\mathcal{T})$ agrees with the Poisson product on $R/(T)$ so it suffices to prove that $\{J(\mathcal{M}), J(\mathcal{M})\} \subset J(\mathcal{M})$.

Remark. The proof that the element $\mathcal{T}$ in the ring A satisfies (1)–(4) is easy. For example, we can prove (3) as follows. Take $\alpha \in A$ and assume that $\mathcal{T}\alpha = 0$ in the ring A. Now α is the image of some x in R and $\mathcal{T}\alpha = 0$ means that $Tx \in (T^2)$ in the ring R. So $Tx = T^2u$ for some $u \in R$ and then $T(x - Tu) = 0$ gives $x - Tu \in (T)$ so that $x \in (T)$ and it follows that $\alpha \in (\mathcal{T})$ as required.

Also, if M is a left R-module for which $\{u \in M : Tu = 0\} \subset TM$ then we take the A-module $\mathcal{M} = M/T^2M$ and the same argument as above shows that $\{u \in \mathcal{M} : \mathcal{T}u = 0\} \subset \mathcal{T}\mathcal{M}$.

Summing up, it suffices to prove Theorem 1.3 under the special assumption that $T^2 = 0$. Observe that if $T^2 = 0$ then Condition (4) implies that T is a central element in the ring.

2. The use of localizations. From now on we assume that $T^2 = 0$ and we consider a left R-module M for which $\{u \in M : Tu = 0\} \subset TM$ and M/TM is a finitely generated module over the Noetherian ring $R/(T)$. Well-known formulas from commutative algebra show that $J(M) = \mathcal{P}_1 \cap \cdots \cap \mathcal{P}_s$ where $\mathcal{P}_1, \ldots, \mathcal{P}_s$ are the minimal prime ideals in the support of the $R/(T)$-module M/TM.

In other words, if $\operatorname{Supp}(M/TM) = \{\mathcal{P} \in \operatorname{Spec}(R/(T)) :$ the localizations $(M/TM)_{\mathcal{P}} \neq 0\}$ then $(M/TM)_{\mathcal{P}_v} \neq 0$ for each minimal prime divisor $\mathcal{P}_v$ of the radical ideal $J(M)$ and, in addition, the $(R/(T))_{\mathcal{P}_v}$-module $(M/TM)_{\mathcal{P}_v}$ is zero-dimensional, i.e., it is annihilated by some power of the maximal ideal $\mathcal{P}_v(R/(T))_{\mathcal{P}_v}$ in the local ring $(R/(T))_{\mathcal{P}_v}$.

We want to prove that $\{J(M), J(M)\} \subset J(M)$ and obviously this inclusion follows if $\{\mathcal{P}_v, \mathcal{P}_v\} \subset \mathcal{P}_v$ for each minimal prime divisor of $J(M)$.

So take $\mathcal{P} = \mathcal{P}_1$ say and let us try to prove $\{\mathcal{P}, \mathcal{P}\} \subset \mathcal{P}$. To get this we consider the set $S = \{x \in R :$ its image $\overline{x} \in R/(T)$ does not belong to the prime ideal $\mathcal{P}\}$.

Then S is a multiplicative subset of R. Since $T^2 = 0$ and $R/(T)$ is commutative it follows that the universal S-inverting ring R_S is equal to the 2-sided Ore ring $S^{-1}R = RS^{-1}$. We refer to §6 for this fact. We have a canonical mapping $R \to R_S$ and we let $\mathcal{T}$ be the image of T. So then $\mathcal{T} \in R_S$ and we can prove that R_S is a $\mathcal{T}$-ring, i.e., the element $\mathcal{T}$ in the ring R_S satisfies (1)–(4). This is left as an exercise to the reader. It uses $R_S = S^{-1}R = RS^{-1}$ and the flatness of the R-module R_S. Of course, we also use the standard properties of Ore extensions as explained in §6 to prove that $\mathcal{T}$ satisfies Condition (4), for example.

Concerning the ring $R_S/(\mathcal{T})$ we see that it is commutative and it equals the local ring $(R/(T))_{\mathcal{P}}$. Hence $R_S/(\mathcal{T})$ is a local Noetherian ring. The Poisson product on $R_S/(\mathcal{T})$ is defined exactly as in §1.1 using the $\mathcal{T}$-ring R_S. It is not difficult to prove that this Poisson product on $R_S/(\mathcal{T})$ via the isomorphism $R_S/(\mathcal{T}) \cong (R/(T))_{\mathcal{P}}$ coincides with the extension of $\{\ ,\ \}$ on $R/(T)$ to its localization $(R/(T))_{\mathcal{P}}$.

In other words, if $\delta \in R/(T) - \mathcal{P}$ so that the inverse δ^{-1} exists in the local ring $(R/(T))_{\mathcal{P}}$ and if $\alpha \in R/(T)$, then $\{\delta^{-1}, \alpha\}$ equals $-\delta^{-2}\{\delta, \alpha\}$ and so on.

Finally, if M is a left R-module we get the left R_S-module $M_S = R_S \otimes_R M$ and here $M_S/\mathcal{T}M_S = (M/TM)_{\mathcal{P}}$, and using the flatness of the R-module R_S, the inclusion $\{u \in M : Tu = 0\} \subset TM$ implies that $\{w \in M_S : \mathcal{T}w = 0\} \subset \mathcal{T}M_S$ easily follows.

Summing up, using the localization with respect to the multiplicative subset S we can work with the R_S-module M_S, and in order to prove that $\{\mathcal{P}, \mathcal{P}\} \subset \mathcal{P}$ holds it suffices to prove that $\{\mathfrak{m}, \mathfrak{m}\} \subset \mathfrak{m}$ where $\mathfrak{m} = \mathcal{P}(R/(T))_{\mathcal{P}}$ is the maximal ideal in the local ring $(R/(T))_{\mathcal{P}}$.

Observe also that $\mathcal{P}$ was a minimal prime divisor of $J(M)$ so that $\mathfrak{m}$ equals $J(M_S)$, i.e., $M_S/\mathcal{T}M_S \cong (M/TM)_{\mathcal{P}}$ is a zero-dimensional module over the local ring $R_S/\mathcal{T}R_S \cong (R/(T))_{\mathcal{P}}$.

3. An Artinian reduction. It remains to prove Theorem 1.3 under the following additional assumptions: $R/(T)$ is a local Noetherian ring with its maximal ideal $\mathfrak{m}$ and M is a left R-module such that $J(M) = \mathfrak{m}$ and $\{u \in M : Tu = 0\} \subset TM$. Then we must prove that $\{\mathfrak{m}, \mathfrak{m}\} \subset \mathfrak{m}$.

To prove this we consider $\mathcal{M} = \{x \in R : \overline{x} \in \mathfrak{m}\}$ and observe that $\mathcal{M}$ is a two-sided ideal in the ring R. Using the definition of the Poisson product on $R/(T)$ we find

3.1. LEMMA. *If* $[\mathcal{M}, \mathcal{M}] = \{xy - yx :$ *both* x *and* $y \in \mathcal{M}\} \subset T\mathcal{M}$ *then* $\{\mathfrak{m}, \mathfrak{m}\} \subset \mathfrak{m}$.

Let us now consider the R-module M above for which $J(M) = \mathfrak{m}$. Here M/TM is a finitely generated and zero-dimensional module over the local Noetherian ring $R/(T)$, so there exists some integer s such that $\mathfrak{m}^s(M/TM) = 0$. Returning to the R-module M this means that $\mathcal{M}^s M \subset TM$, and since $T^2 = 0$ and T is a central element in the ring R we find that $\mathcal{M}^{2s}M = 0$. If we put $\mathcal{A} = R/\mathcal{M}^{2s}$ then we can consider M as a left $\mathcal{A}$-module. Also, if $\overline{\mathcal{M}}$ is the 2-sided ideal in the ring A which is the image of $\mathcal{M}$ and if $\mathcal{T}$ is the image of T, then $[\overline{\mathcal{M}}, \overline{\mathcal{M}}] \subset \mathcal{T}\overline{\mathcal{M}} \Rightarrow [\mathcal{M}, \mathcal{M}] \subset T\mathcal{M} + \mathcal{M}^{2s}$.

Observe here that we can use any $s \geq s_0$ as soon as $\mathfrak{m}^{s_0}(M/TM) = 0$. We want to prove that $[\mathcal{M}, \mathcal{M}] \subset T\mathcal{M}$. If we have proved that $[\mathcal{M}, \mathcal{M}] \subset T\mathcal{M} + \mathcal{M}^{2s}$ for all $s \geq s_0$, the required inclusion $[\mathcal{M}, \mathcal{M}] \subset T\mathcal{M}$ follows from

3.2. LEMMA. $T\mathcal{M} = \bigcap_{s \geq 1}(T\mathcal{M} + \mathcal{M}^{2s})$.

Proof. $\mathcal{M}/T\mathcal{M}$ is a finitely generated $R/(T)$-module. Indeed, first $\mathfrak{m}$ is a finitely generated $R/(T)$-module and if $u_1, \dots, u_s$ is a finite set of elements in $\mathcal{M}$ such that their images $\overline{u}_1, \dots, \overline{u}_s$ in $R/(T)$ generate the ideal $\mathfrak{m}$, then their images in $\mathcal{M}/T\mathcal{M}$ are generators of this $R/(T)$-module. Now Krull's intersection theorem is applied to the Noetherian $R/(T)$-module $\mathcal{M}/T\mathcal{M}$, i.e., $\bigcap_{v \geq 1} \mathfrak{m}^v(\mathcal{M}/T\mathcal{M}) = 0$ holds and this gives Lemma 3.2, as the reader may verify.

Summing up, if we can prove that $[\mathcal{M}, \mathcal{M}] \subset T\mathcal{M} + \mathcal{M}^{2s}$ for all $s \geq s_0$, then Theorem 1.3 is proved.

To prove this we choose s_0 so that $\mathfrak{m}^{s_0}(M/TM) = 0$ and if $s \geq s_0$ we introduce the ring $A = R/\mathcal{M}^{2s}$ and then the left R-module M satisfies $\mathcal{M}^{2s}M = 0$ and hence it can be identified with a left A-module. Keeping $s \geq s_0$ fixed we consider the image of T in $R/\mathcal{M}^{2s}$ and denote it by $\mathcal{T}$. Then $A/(\mathcal{T}) = (R/(T))/\mathfrak{m}^{2s}$ is a local Artinian ring and the assumption that $\{u \in M : Tu = 0\} \subset TM$ gives $\{u \in M : \mathcal{T}u = 0\} \subset \mathcal{T}M$ when we treat M as a left A-module.

At this stage we conclude that Theorem 1.3 follows from the general result below.

3.3. PROPOSITION. *Let A be a Q-algebra which contains a central element $\mathcal{T}$ such that $\mathcal{T}^2 = 0$ and $A/(\mathcal{T})$ is a local Artinian ring. Assume that there exists a left A-module M which is not identically zero and for which $M/\mathcal{T}M$ is a finitely generated $A/(\mathcal{T})$-module and $\{u \in M : \mathcal{T}u = 0\} \subset \mathcal{T}M$. Then*

$[\mathcal{M}, \mathcal{M}] \subset \mathcal{T}\mathcal{M}$ *where* $\mathcal{M} = \{\alpha \in A$: *its image* $\overline{\alpha}$ *in* $A/(\mathcal{T})$ *belongs to the maximal ideal of the local ring* $A/(\mathcal{T})\}$.

Remark. Observe that we do not refer to a Poisson product on the ring $A/(\mathcal{T})$. In fact, we do not assume Condition (3) for the element, so a Poisson product need not exist on $A/(\mathcal{T})$. During the passage from a T-ring R to $A = R/\mathcal{M}^{2s}$ above it may occur that Condition (3) no longer holds for the image of T in the ring A. However, when we studied the R-module M for which M/TM is zero-dimensional, the assumption that $\{u \in M : Tu = 0\} \subset TM$ was inherited by the A-module M when $s \geq s_0$.

We prove Proposition 3.3 in §4. Observe that the inclusion $[\mathcal{M}, \mathcal{M}] \subset \mathcal{T}\mathcal{M}$ is derived from the existence of an A-module M which satisfies the two conditions above, i.e., $M/\mathcal{T}M$ is a finitely generated $A/(\mathcal{T})$-module and $\{u \in M : \mathcal{T}u = 0\} \subset \mathcal{T}M$. It is illuminating to see an example where the ring A contains a central element $\mathcal{T}$ for which $\mathcal{T}^2 = 0$ and $A/(\mathcal{T})$ is a local Artinian ring and yet $[\mathcal{M}, \mathcal{M}] \subset \mathcal{T}\mathcal{M}$ does not hold.

3.4. *Example.* Let $A = Q\varepsilon + Q\xi + Q\eta + Q\xi\eta + Q\delta$ be a 5-dimensional Q-algebra where the multiplications satisfy: $\xi^2 = \eta^2 = \xi\delta = \delta\xi = \eta\delta = \delta\eta = \delta^2 = 0$ while $\eta\xi = \xi\eta + \delta$. Then δ is a central element in the ring A and $A/(\delta)$ is a commutative local Artinian ring. Both ξ and η belong to $\mathcal{M}$ and $\eta\xi - \xi\eta = \delta$ is outside $\delta\mathcal{M}$. So for the ring A we do not get $[\mathcal{M}, \mathcal{M}] \subset \delta\mathcal{M}$ and the proof of Proposition 3.3 shows that there does not exist a nonzero A-module M for which $\{u \in M : \delta u = 0\} \subset \delta M$ and $M/\delta M$ is a finitely generated $A/(\delta)$-module.

4. Proof of Proposition 3.3. Let A and M be as in Proposition 3.3. If x and $y \in \mathcal{M}$ we consider the commutator $xy - yx$ which belongs to the two-sided ideal $(\mathcal{T})$ since $A/(\mathcal{T})$ is a commutative ring. Hence $xy - yx = \mathcal{T}u$ for some $u \in A$ and we want to prove that u belongs to the two-sided maximal ideal $\mathcal{M}$. Observe that u is not uniquely determined by x and y. However, the condition that $u \in \mathcal{M}$ is intrinsic, i.e., if $xy - yx = \mathcal{T}u'$ for another element u' then $\mathcal{T}(u - u') = 0$, and since we assume that $\mathcal{T} \neq 0$, it follows that $u - u'$ cannot be invertible in the ring A and hence $u \in \mathcal{M} \Leftrightarrow u' \in \mathcal{M}$.

So let $xy - yx = \mathcal{T}u$ and, in order to prove that $u \in \mathcal{M}$, we are going to use the A-module M.

First we need some preliminaries before $u \in \mathcal{M}$ is proved.

4.1. *The field* K. $A/(\mathcal{T})$ is a local Artinian ring and, since A is a Q-algebra, it follows that $A/(\mathcal{T})$ is a Q-algebra, and similarly we find that if $\mathfrak{m}$ is the maximal ideal in the local Artinian ring $A/(\mathcal{T})$, which we denote by $\overline{A}$, then $\overline{A}/\mathfrak{m}$ is a field of characteristic zero.

This implies that $\overline{A}/\mathfrak{m}$ has a representative field, i.e., there exists a subfield K of $\overline{A}$ so that $K \oplus \mathfrak{m} = \overline{A}$.

Let us now consider $M/\mathcal{T}M$, which by assumption is a finitely generated $A/(\mathcal{T})$-module. Observe that $\overline{A} = A/(\mathcal{T})$ is a finite-dimensional K-algebra, which implies that $M/\mathcal{T}M$ is a finite-dimensional vector space over the field

K. So if $\alpha \in A$ and if $\overline{\alpha}$ is its image in $\overline{A}$ then we get the K-linear mapping $M/\mathcal{T}M \xrightarrow{\overline{\alpha}} M/\mathcal{T}M$ derived from the multiplication by α on the given left A-module M.

In particular, if x and $y \in \mathcal{M}$ and if $xy - yx = \mathcal{T}u$, we find $\overline{u}$ in $\mathrm{Hom}_K(M/\mathcal{T}M, M/\mathcal{T}M)$ and we can compute the usual trace of this K-linear mapping. In $\overline{A}$ we have $\overline{u} = \lambda + w$ with $\lambda \in K$ and $w \in \mathfrak{m}$. Here some power of the maximal ideal $\mathfrak{m}$ is zero. In particular, the K-linear operator which w determines on the K-space $M/\mathcal{T}M$ is nilpotent so that $\mathrm{Trace}(w) = 0$ and hence $\mathrm{Trace}(\overline{u}) = \lambda m$ where $m = \dim_K(M/\mathcal{T}M)$.

We conclude that $\overline{u} \in \mathfrak{m}$ if $\mathrm{Trace}(\overline{u}) = 0$. Since $\overline{u} \in \mathfrak{m} \Rightarrow u \in \mathcal{M}$ it remains to prove that $\mathrm{Trace}(\overline{u}) = 0$ and this is achieved below.

4.2. *Some matrix formulas.* On the K-space $M/\mathcal{T}M$ the K-linear operators $\overline{\alpha}$ with $\overline{\alpha} \in \mathfrak{m}$ are pairwise nilpotent operators. Hence we can choose a basis $\varepsilon_1, \dots, \varepsilon_m$ of the m-dimensional K-space $M/\mathcal{T}M$ such that if $\overline{\alpha} \in \mathfrak{m}$ then $\overline{\alpha}\varepsilon_j \in K\varepsilon_1 + \cdots + K\varepsilon_{j-1}$ for all $j \geq 2$ and $\overline{\alpha}\varepsilon_1 = 0$. So it means that the matrix $M(\overline{\alpha})$ which represents $\alpha\varepsilon$ is strictly upper triangular, i.e., $\overline{\alpha}\varepsilon_j = a_{1j}\varepsilon_1 + \cdots + a_{mj}\varepsilon_m$ with $a_{vj} = 0$ for all $v \geq j$.

Let us take $x \in \mathcal{M}$ so that $\overline{x} \in \mathfrak{m}$ and now $\overline{x}\varepsilon = M(\overline{x})\varepsilon$. In the given A-module M we choose $e_1, \dots, e_m$ so that the images $\overline{e}_j = \varepsilon_j$ in $M/\mathcal{T}M$.

In the ring A we choose elements a_{vj} so that $M(\overline{x}) =$ the matrix $(\overline{a}_{vj})$ and here we can take $a_{vj} = 0$ when $v \geq j$.

It follows that $xe_j - (a_{1j}e_1 + \cdots + a_{mj}e_m) \in \mathcal{T}M$ for all j. Since $\mathcal{T}^2 = 0$ and $M = Ae_1 + \cdots + Ae_m + \mathcal{T}M$ we get $\mathcal{T}M = \mathcal{T}Ae_1 + \cdots + \mathcal{T}Ae_m$ and hence we can choose an A-valued matrix (c_{vj}) so that

$$xe_j = \sum a_{vj}e_v + \sum c_{vj}e_v \text{ hold in } M \text{ for all } 1 \leq j \leq m.$$

With $\mathcal{A} = (a_{vj})$ and $\mathcal{C} = (c_{vj})$ these equations are given in a matrix form as $x\mathbf{e} = \mathcal{A}\mathbf{e} + \mathcal{T}\mathcal{C}\mathbf{e}$.

Starting from another element $y \in \mathcal{M}$ we find $y\mathbf{e} = \mathcal{B}\mathbf{e} + \mathcal{T}\mathcal{D}\mathbf{e}$ in the same way. Here $\mathcal{B} = (b_{vj})$ has $b_{vj} = 0$ for all $v \geq j$, i.e., the matrices $\mathcal{A}$ and $\mathcal{B}$ above are strictly upper triangular.

4.3. *A formula for $xy\mathbf{e}$.* Here $(xy)\mathbf{e} = x(\mathcal{B}\mathbf{e} + \mathcal{T}\mathcal{D}\mathbf{e}) = [x, \mathcal{B}]\mathbf{e} + \mathcal{B}x\mathbf{e} + \mathcal{T}[x, \mathcal{D}]\mathbf{e} + \mathcal{T}\mathcal{D}x\mathbf{e}$ where $[x, \mathcal{B}]$ is the matrix whose entries are the commutators $xb_{vj} - b_{vj}x$ and similarly for $[x, \mathcal{D}]$.

Since $\mathcal{T}^2 = 0$ and the entries $xd_{vj} - d_{vj}x \in \mathcal{T}$ we have $\mathcal{T}[x, \mathcal{D}] = 0$.

We conclude that $(xy)\mathbf{e} = \mathcal{B}\mathcal{A}\mathbf{e} + \mathcal{T}\mathcal{B}\mathcal{C}\mathbf{e} + \mathcal{T}\mathcal{D}\mathcal{A}\mathbf{e} + [x, \mathcal{B}]\mathbf{e}$.

A similar formula is found for $(yx)\mathbf{e}$ and it leads to

4.4. *Formula.* $(xy - yx)\mathbf{e} = [x, \mathcal{B}]\mathbf{e} + [y, \mathcal{A}]\mathbf{e} + (\mathcal{B}\mathcal{A} - \mathcal{A}\mathcal{B})\mathbf{e} + \mathcal{T}(\mathcal{D}\mathcal{A} - \mathcal{A}\mathcal{D})\mathbf{e} + \mathcal{T}(\mathcal{B}\mathcal{C} - \mathcal{C}\mathcal{B})\mathbf{e}$.

Here $xb_{vj} - b_{vj}x \in (\mathcal{T})$ for all v and j and similarly the entries of $[y, \mathcal{A}]$ all belong to $(\mathcal{T})$. Since both $\mathcal{B}$ and $\mathcal{A}$ are strictly upper triangular we can write $[x, \mathcal{B}] = \mathcal{T}\chi$ and $[y, \mathcal{A}] = \mathcal{T}\Psi$ where both χ and Ψ are strictly upper triangular.

Next, the images $\overline{\mathcal{A}}$ and $\overline{\mathcal{B}}$ represent $M(\overline{x})$ and $M(\overline{y})$ and here $\overline{x}$ and $\overline{y}$ commute so that $(\mathcal{B}\mathcal{A} - \mathcal{A}\mathcal{B})\mathbf{e} = \mathcal{T}\Phi\mathbf{e}$ where we again can choose Φ to be strictly upper triangular since both $\mathcal{A}$ and $\mathcal{B}$ are strictly upper triangular.

Summing up, $(xy - yx)\mathbf{e} = \mathcal{T}(\chi + \Psi + \Phi)\mathbf{e} + \mathcal{T}[\mathcal{D}, \mathcal{A}]\mathbf{e} + \mathcal{T}[\mathcal{B}, \mathcal{C}]\mathbf{e}$ holds. If $xy - yx = \mathcal{T}u$ in the ring A we find that $\mathcal{T}(u\mathbf{e} - (\Lambda + [\mathcal{D}, \mathcal{A}] + [\mathcal{B}, \mathcal{C}]))\mathbf{e} = 0$ where $\Lambda = \chi + \Psi + \Phi$ is strictly upper triangular.

Now we use that $\{\xi \in M : \mathcal{T}\xi = 0\} \subset \mathcal{T}M$ and conclude that the following holds.

4.5. *Formula.* $(u - (\Lambda + [\mathcal{D}, \mathcal{A}] + [\mathcal{B}, \mathcal{C}])\mathbf{e} \in \mathcal{T}M$.

Passing to $M/\mathcal{T}M$ it means that $\overline{u}\varepsilon = \overline{\Lambda}\varepsilon + [\overline{\mathcal{D}}, \overline{\mathcal{A}}]\varepsilon + [\overline{\mathcal{B}}, \overline{\mathcal{C}}]\varepsilon$ and here $\overline{\Lambda}$ is strictly upper triangular so its trace is zero. Also, $\text{Trace}([\overline{\mathcal{D}}, \overline{\mathcal{A}}]) = \text{Trace}([\overline{\mathcal{B}}, \overline{\mathcal{C}}]) = 0$ since the trace of any commutator of K-valued matrices is zero.

Hence $\text{Trace}(\overline{u}) = 0$ follows, which proves Proposition 3.3.

5. The case of filtered rings. To any filtered ring R we can associate a T-ring and then apply Theorem 1.3. To be precise, let $\Sigma_{v-1} \subset \Sigma_v \subset \Sigma_{v+1}$ be a filtration on R such that $\text{gr}(R) = \oplus\Sigma_v/\Sigma_{v-1}$ is commutative. Here the multiplicative unit 1 belongs to Σ_0.

Now we can consider the ring $\mathcal{R} = \oplus\Sigma_v$ where each Σ_v appears as a direct summand. Of course, if $x \in \Sigma_k$ and if $y \in \Sigma_v$ then we have $xy \in \Sigma_{k+v}$ in the given ring R and this is used to define products between pairs of homogeneous elements in $\oplus\Sigma_v$ and it extends to give a ring structure on $\mathcal{R}$ by addition.

5.1. *Remark.* If T is a new indeterminate we consider the ring $R[T, T^{-1}]$ of finite Laurent series with coefficients in R. It contains the subring whose elements are of the form $\oplus r_v T^v$ where $r_v \in \Sigma_v$ for all v. This subring of $R[T, T^{-1}]$ is isomorphic to $\mathcal{R}$. The element T is homogeneous of degree 1 in $\mathcal{R} \subset R[T, T^{-1}]$ and $\mathcal{R}/(T)$ is canonically isomorphic to the commutative ring $\text{gr}(R)$.

Observe that T is a nonzero divisor on $\mathcal{R}$, so $\mathcal{R}$ is a T-ring and if we define the Poisson product on $\mathcal{R}/(T)$ following the construction in 1.1., then it agrees with the Poisson product which we construct on $\text{gr}(R)$ using the following procedure.

5.2. *The Poisson product on* $\text{gr}(R)$. Let σ_k be the canonical mapping from Σ_k to Σ_k/Σ_{k-1}. If $x \in \Sigma_k$ and if $y \in \Sigma_v$ where k and v are some integers, then the commutator $xy - yx$ belongs to Σ_{k+v-1} because $\text{gr}(R)$ is commutative and we define $\{\sigma_k(x), \sigma_v(y)\} = \sigma_{k+v-1}(xy - yx)$. This gives a Poisson product on $\text{gr}(R)$ which coincides with $\{\ ,\ \}$ on $\mathcal{R}/(T)$ when we identify $\text{gr}(R)$ with the T-ring $\mathcal{R}$.

Applying Theorem 1.3, we conclude that if M is a left R-module and if Γ is a filtration on M such that $\text{gr}_\Gamma(M) = \oplus\Gamma_v/\Gamma_{v-1}$ is a finitely generated $\text{gr}(R)$-module, then the radical of $I_\Gamma(M)$ is closed under the Poisson product. Of course, we have to assume that R is a Q-algebra and that $\text{gr}(R)$ is a commutative Noetherian ring for this to be true.

In the case when R is left and right Noetherian and M is a finitely generated R-module, we can choose a good filtration Γ and conclude that $\{\text{Ch}(M), \text{Ch}(M)\} \subset \text{Ch}(M)$ where $\text{Ch}(M)$ is the characteristic ideal which was found in §6 of Part I.

6. Noncommutative localizations. In §2 we used localizations of a noncommutative ring. This is perhaps less well known as compared with localizations of commutative rings, so we insert some remarks following P. M. Cohn's book [8] where Chapter 1 contains enough material for the applications in §2.

6.1. *The ring R_S.* Let R be a ring and let S be a multiplicative subset of R, i.e., s and $t \in S \Rightarrow st \in S$. By an S-inverting pair (φ, A) we mean a ring A and a ring homomorphism $\varphi: R \to A$ such that $\varphi(s)$ are invertible in A for all $s \in S$. Among all S-inverting pairs there exists a "universal pair" (φ_0, R_S) such that whenever (φ, A) is S-inverting, then there exists a ring homomorphism $f: R_S \to A$ with $\varphi = f \circ \varphi_0$.

6.2. *How to construct R_S.* First the ring R is represented as the quotient of a free associative ring $\mathcal{R}$, i.e., $\mathcal{R}$ is a free module over the ring of integers Z formed by ordered monomials in a family of generators $\{\varepsilon_\alpha\}$ where we have chosen from the start a family $\{r_\alpha\}$ of elements in the ring R such that they generate R as a ring.

Then $R \cong \mathcal{R}/J$ where J is the kernel under the ring homomorphism $\mathcal{R} \to R$ which sends ε_α into r_α for all indices α.

Let us then consider the multiplicative subset S where we can assume from the start that some family, say $\{\varepsilon_\beta: \beta \in \mathcal{B}\}$, corresponds to elements in the set S. We introduce a new copy of the set S, say $\{s_\beta : \beta \in \mathcal{B}\}$, and find the free ring $\tilde{\mathcal{R}}$ with generators $\{\varepsilon_\alpha\} \cup \{s_\beta\}$. In $\tilde{\mathcal{R}}$ we divide out by the 2-sided ideal generated by J and by $\{\varepsilon_\beta s_\beta - 1 : \beta \in \mathcal{B}\}$, and if we call this ideal $\tilde{J}$ then $\tilde{\mathcal{R}}/\tilde{J} = R_S$.

6.3. *Remark.* The reader may consult P. M. Cohn's book [8] for the proof that R_S solves the universal problem among the S-inverting pairs. We call R_S the S-inverting ring.

6.4. *Ore's construction.* In special cases we can find R_S in a more direct way. Namely, we say that S satisfies the *right Ore condition* if the following is true:

(1) $aS \cap sR$ is a nonempty set for any pair $a \in R$ and $s \in S$;

(2) If $sa = 0$, then $\exists t \in S$ with $ta = 0$.

If the multiplicative subset S satisfies the right Ore condition, one proves that R_S is the ring of right fractions. To be precise, we find $R_S = RS^{-1}$ where the elements in RS^{-1} have the form as^{-1} with $a \in R$ and $s \in S$. Here $as^{-1} = bt^{-1}$ with $a, b \in R$ and s and $t \in S$ if and only if there exist a' and b' in R so that $sa' = tb' \in S$ and $aa' = bb'$.

Again we refer to Cohn [8] for the proof that $R_S = RS^{-1}$ when S satisfies the right Ore condition and also for a detailed discussion about the ring RS^{-1} which, in particular, shows that it is flat as a left R-module.

In addition, we can impose a similar *left Ore condition* and find the ring $S^{-1}R$ of left fractions. If S satisfies both the left and the right Ore condition then $S^{-1}R = RS^{-1} = R_S$.

6.5. *Example.* Let R be a T-ring where $T^2 = 0$. Then any multiplicative subset S of R satisfies the two Ore conditions. To see this we take $s \in S$ and $a \in R$. Now $sa - as \in (T)$ and since $T^2 = 0$ we find that the two-fold commutator

$s(sa - as) - (sa - as)s = 0$. This gives $as^2 = s(2as - sa)$ and here $s^2 \in S$, which first proves that $aS \cap sR$ is nonempty. Also, if $sa = 0$ we get $at = 0$ with $t = s^2 \in S$, i.e., Condition (2) above holds. In the same way we can verify the left Ore condition.

III. Rings of differential operators

Summary. This part does not offer detailed proofs since it would take us too far. The theory about some classical rings of differential operators, such as the Weyl algebra or the ring of differential operators with analytic coefficients over a real (or complex) analytic manifold, has developed during the last 15 years. Here the ring theory has been motivated by questions from analysis or geometry. For example, one can use ring-theoretic results about the Weyl algebra $A_n(\mathbb{C})$ to construct fundamental solutions to differential operators. In geometry the theory about holonomic modules has culminated in the *Riemann–Hilbert correspondence*, which gives an isomorphism between the abelian category of holonomic sheaves with regular singularities and constructible sheaf complexes satisfying the perversity condition. See, for example, Brylinski's work [7] for this.

Here we only present a brief introduction where the preceding material is applied to the Weyl algebra $A_n(\mathbb{C})$. For example, in §1 we discuss two filtrations, the $\mathcal{T}$-filtration and the Σ-filtration. The $\mathcal{T}$-filtration was used by I. N. Bernstein to get proofs of some basic results about A_n. The most notable is that $0 \leq j(M) \leq n$ holds for any finitely generated A_n-module. Both the $\mathcal{T}$- and the Σ-filtrations are discrete, so the results from §4 in Part I can be used. Since the associated graded rings in both cases give the polynomial ring in $2n$ variables we find, in particular, that A_n is a *Noetherian Gorenstein ring* and we can also compute its global homological dimension which turns out to be n. Here the inequality $\text{gl.dim}(A_n) \leq n$ is remarkable and it is the reason why the family of *holonomic* A_n-modules contains remarkably "large modules". We discuss the main example in detail. It arises when we take any polynomial $P(x)$ in $\mathbb{C}[x] = \mathbb{C}[x_1, \cdots, x_n]$ and consider $\mathbb{C}[x, P^{-1}]$ as a left A_n-module. We mention the various proofs which show that this A_n-module is holonomic.

In addition, we mention some results which go beyond the material in Part I, i.e., here we need proofs which use the particular structure of the filtered ring A_n. For example, the characteristic ideal of $\mathbb{C}[x, P^{-1}]$ can be understood but the proof uses *resolution of singularities*, so it lies deep.

In §2 we consider the "Fuchsian filtration" with respect to a hyperplane. To be precise, we take a new variable t and work with A_{n+1}, and it is filtered when we let $\text{ord}(t) = -1$ and $\text{ord}(D_t) = +1$ while x_v and D_{x_v} have order zero for all v. This filtration does not satisfy the closure condition, but it satisfies the comparison condition and it has a natural geometric significance based upon *Deligne's constructions of vanishing cycles.*

We offer some illuminating computations to reveal a bit about this. In particular, we are able to *define* the famous b-function—here we call it the Bernstein–Sato polynomial in honor of I. N. Bernstein and M. Sato who have both contributed to its existence in the present general setup.

Finally, in the final section, we give some comments about the material in §1 and §2. For example, we mention J. T. Stafford's striking example of *nonholonomic A_n-modules whose Krull dimensions are zero.*

1. The Weyl algebra and its filtrations. Let K be a commutative field of characteristic zero and $A_n(K)$ the ring of K-linear differential operators with polynomial coefficients. This means that $A_n(K)$ is a K-algebra with generators $x_1, \ldots, x_n, D_1, \ldots, D_n$ where $D_v = \partial/\partial x_v$. An element in $A_n(K)$ is a differential operator, which can be written in the form $\Sigma q_\alpha(x) D^\alpha$ where Σ extends over finitely many multi-indices and $D^\alpha = D_1^{\alpha_1} \cdots D_n^{\alpha_n}$. Here $q_\alpha(x) \in K[x_1, \ldots, x_n]$.

Remark. The n-tuple $x_1, \ldots, x_n$ commute pairwise and similarly $D_1, \ldots, D_n$ commute pairwise. On the other hand, the commutators $[D_j, x_j] = D_j x_j - x_j D_j = 1$ for all j.

We refer to [4: Chapter 1] for a detailed presentation of the ring $A_n(K)$. From now on we assume that the field K is the complex field $\mathbb{C}$ since we want to express some results in a more geometric way via the Nullstellensatz.

1.1. *The $\mathcal{T}$-filtration.* First, $A_n(\mathbb{C})$ is a vector space over $\mathbb{C}$ where $\{x^\beta D^\alpha : \beta$ and $\alpha \geq 0\}$ is a basis. If $\mathcal{T}_v = \oplus \mathbb{C} x^\beta D^\alpha : |\beta| + |\alpha| \leq v$, then $\mathcal{T}_0 \subset \mathcal{T}_1 \subset \cdots$ is a filtration on the ring $A_n(\mathbb{C})$ and one proves easily that $\operatorname{gr}(A_n)$ is commutative. In fact, $\operatorname{gr}(A_n) = \mathbb{C}[\overline{x}, \overline{D}]$ is the polynomial ring in $2n$ variables where $\overline{x}_v$ and $\overline{D}_v$ are images of x_v and D_v in $\mathcal{T}_1/\mathcal{T}_0$.

Since a polynomial ring is a regular commutative Noetherian ring, it is, in particular, Gorenstein. So we find that A_n is a Noetherian Gorenstein ring. In particular, we can study the $\mathcal{T}$-filtrations on finitely generated A_n-modules. Observe that $\mathcal{T}_v = 0$ when $v < 0$ implies that this filtration satisfies the closure condition. So if M is a finitely generated A_n-module, we can use the results from §4 in Part I to get $j(M) = j(\operatorname{gr}_\Gamma(M))$ where Γ is some good filtration on M. In [4: Chapter 1] we have reviewed I. N. Bernstein's work [1] where the $\mathcal{T}$-filtration on A_n was studied. In particular, one finds the inequality $\delta(M) \geq n$ by quite elementary methods, i.e., a clever induction combined with a counting of multiplicities. Using the results from Part I, this gives $0 \leq j(M) \leq n$ for any finitely generated A_n-module.

1.2. *The Σ-filtration.* Another filtration on A_n is used when we consider the usual order of differential operators. So here we put $\Sigma_v = \{Q(x, D) = \sum q_\alpha(x) D^\alpha : q_\alpha = 0 \text{ if } |\alpha| > v\}$. Using $\Sigma_0 \subset \Sigma_1 \subset \cdots$ one finds that the graded ring $\oplus \Sigma_v/\Sigma_{v-1}$ again is isomorphic to a polynomial ring in $2n$ variables, namely, $\mathbb{C}[x, \xi]$ where ξ_v are the images of D_v in Σ_1/Σ_0.

The Σ-filtration is more significant than the $\mathcal{T}$-filtration. If we take a finitely generated A_n-module M and choose a good filtration on it, this time with respect to the Σ-filtration, then we get the $\mathbb{C}[x,\xi]$-module $\mathrm{gr}_\Gamma(M)$ and the characteristic ideal $J(M)$, which becomes an ideal in $\mathbb{C}[x,\xi]$ that is generated by ξ-homogeneous polynomials.

The locus $J(M)^{-1}(0) = \{(x,\xi) \in \mathbb{C}^{2n} : q(x,\xi) = 0 \text{ for all } q \in J(M)\}$ is an algebraic set in $\mathbb{C}^{2n}$ which is ξ-conic. Gabber's integrability theorem in Part II gives $\{J(M), J(M)\} \subset J(M)$. Here the Poisson product as defined in Part II agrees with the usual Poisson product on the $2n$-dimensional (x,ξ)-space which is identified with the cotangent space over $\mathbb{C}^n$. See for example [4: pp. 148–151] for a discussion about this. In particular, it follows that $\dim(J(M)^{-1}(0)) \geq n$ (unless this locus is empty, which only occurs if $J(M) = \mathbb{C}[x,\xi] \Rightarrow M = 0$).

Remark. To be precise, $\{J(M), J(M)\} \subset J(M)$ implies $J(M)^{-1}(0)$ is an involutive subset of $T^\times(\mathbb{C}^n)$, and then an elementary geometric argument proves that every irreducible component of $J(M)$ has dimension $\geq n$. See for example [Björk: Prop. 5.4.5 on page 151].

Summing up, if $0 \neq M$, then $\dim(J(M)^{-1}(0)) \geq n$, and by familiar dimension theory for modules over polynomial rings, this gives that the j-number of the $\mathbb{C}[x,\xi]$-module $\mathrm{gr}_\Gamma(M)$ is $\leq n$ for any good filtration Γ.

Next, we can use the equality in Theorem 4.3 from Part I and get $0 \leq j(M) \leq n$ for any finitely generated A_n-module. In other words, Gabber's integrability theorem and classical geometry are used to get this inequality, which also could be deduced by the $\mathcal{T}$-filtration and the work by I. N. Bernstein.

1.3. *The formula* $\mathrm{gl.dim}(A_n) = n$. The global homological dimension of A_n equals n. This is related to $j(M) \leq n$, i.e., one can easily prove that $j(M) \leq n = \mathrm{gl.dim}(A_n)$. In [26] J.-E. Roos discovered a purely homological proof of the equality $\mathrm{gl.dim}(A_n) = n$, so it offers another way of getting $j(M) \leq n$. The methods from [26] have been extended, also in [4: Chapter 3], to cover other rings of differential operators. See also work by Goodearl in [14] and also [3].

Remark. As mentioned in Part II, Gabber's result was already known by analytic methods in [28] and the equality $\mathrm{gl.dim}(A_n) = n$ was then found by more analysis and it was, for example, already used by M. Kashiwara in his thesis [16].

1.4. *The A_n-module* $\mathbb{C}[x, P^{-1}]$. Let $P(x) \in \mathbb{C}[x]$ and consider the ring of fractions $= \{Q/P^v : v \geq 0 \text{ and } Q \in \mathbb{C}[x]\}$, which is a left A_n-module. It is not obvious why it is finitely generated. However, using the $\mathcal{T}$-filtration and the inequality $0 \leq j(M) \leq n$ for finitely generated A_n-modules combined with a counting with multiplicities, one can prove that $\mathbb{C}[x, P^{-1}]$ is even a *holonomic* A_n-module, i.e., its j-number takes the maximal value $n = \mathrm{gl.dim}(A_n)$. See also §1.24 in Part I.

We refer to [4: Chapter 1] or [2] for the detailed proof. In §2 we discuss *another proof* which uses results from §4 in Part I and hence offers a "self-contained proof" when material from Parts I and II are accepted.

1.5. *The position of* $J(\mathbb{C}[x, P^{-1}])$. Consider this characteristic ideal and let $V(P)$ denote its locus. Hence $V(P)$ is a ξ-conic algebraic subset in the $2n$-dimensional (x, ξ)-space. Its dimension is n. It is possible to compute $V(P)$ and the result is the next theorem.

1.6. THEOREM. $V(P) = \{(x, \xi) : \xi = 0$ *or* $0 \neq \xi = \lim \lambda_v \operatorname{grad}(P)(x_v)$ *for some sequence* $x_v \to x$ *and some sequence* λ_v *of complex scalars and at the same time* $x \in P^{-1}(0)$, *i.e.*, $P(x) = 0\}$.

Remark. So first $V(P)$ contains the "zero-section" $= \{(x, 0) : x \in \mathbb{C}^n\}$ and, in addition, certain points (x, ξ) enter with $P(x) = 0$ and $0 \neq \xi$. If the algebraic hypersurface $P^{-1}(0)$ is regular then the formula for $V(P)$ is easy to discover. But for a general position of $P^{-1}(0)$ the formula for $V(P)$ is very difficult to get. In fact, the only known proof uses resolution of singularities, i.e., Hironaka's difficult theorem from [15].

In addition to this, some further methods are required. Theorem 1.6 was proved in [30] using the sheaf $\mathcal{E}$ of micro-local differential operators and the theory about determinants of $\mathcal{E}$-valued matrices which was introduced by M. Sato and M. Kashiwara in [29]. See also [5] and [21]. In particular, [21] offers a nice geometric explanation of Theorem 1.6, which also gives formulas for the multiplicities of the module $\mathbb{C}[x, P^{-1}]$ as explained in Remark 5.7.

1.7. *An intersection formula.* Let $L \subset A_n$ be a left ideal and let $\varphi \in \operatorname{Hom}_{A_n}(A_n/L, A_n/L)$. So this means that the left A_n-linear mapping φ is induced by the right multiplication of some $Q \in A_n$ satisfying $LQ \subset L$. If $M = A_n/L$, then $M/\varphi(M) = A_n/(L + A_nQ)$ and we want to relate $J(M)$ with $J(M/\varphi(M))$. Here $J(M) = \sqrt{\sigma(L)}$ and it is obvious that $J(M/\varphi(M))$ contains the radical of the ideal $\sigma(L) + \operatorname{gr}(A_n)\sigma(Q)$, where we are using the Σ-filtration on A_n so that $\operatorname{gr}(A_n)$ is $\mathbb{C}[x, \xi]$.

The question arises whether $\sqrt{\sigma(L + A_nQ)} = \sqrt{\sigma(L) + \operatorname{gr}(A_n)\sigma(Q)}$ holds. This is not true in general, i.e., easy counterexamples can be found. However, one has a positive result if we add the following hypothesis.

Hypothesis. Identify the principal symbol $\sigma(Q)(x, \xi)$ with a function and assume that it is not identically zero on any irreducible component of $J(M)^{-1}(0)$, i.e., $\sigma(Q)$ does not belong to any minimal prime divisor of $J(M)$. Then $J(M/\varphi(M)) = \sqrt{\sigma(L) + \operatorname{gr}(A_n)\sigma(Q)}$, which means that its locus is the intersection between $J(M)^{-1}(0)$ and $\sigma(Q)^{-1}(0)$.

Remark. This result is *easy* if $\sigma(Q)$ is a nonzero divisor on $\operatorname{gr}(A_n)/\sigma(L)$. However, the hypothesis need not imply this because the ideal $\sigma(L)$ may have *imbedded primary components.* So the proof of the equality above, under the hypothesis that $\sigma(Q)$ does not belong to the minimal prime divisors of the radical ideal $\sqrt{\sigma(L)}$, uses methods which go beyond those in §6 of Part I. We refer to [5] for details and mention that it uses a considerable amount of analysis, based upon the sheaf $\mathcal{E}$.

2. Holonomic A_n-modules. By definition a finitely generated A_n-module M is holonomic in $j(M) = n$. If $P(x)$ is a polynomial we have found that $\mathbb{C}[x, P^{-1}]$ is holonomic. It turns out that this can be proved using results from §4 in Part I and another filtration on Weyl algebras.

Namely, take a new coordinate t and use the $n+1$-dimensional (x,t)-space to get the Weyl algebra A_{n-1} where both t and D_t occur. Now we find a filtration on A_{n+1} where $\text{ord}(t) = -1$ and $\text{ord}(D_t) = +1$ while x_v and D_{x_v} have order zero for all v.

2.1. *Formula.* If F^k is the set of elements in A_{n+1} whose order $\leq k$ then $F^k \neq 0$ for all integers. We have $F^0 = \{Q(x, t, D_x, D_t) : Q$ is of the form $\sum_{v \geq 0} Q_v(x, t, D_x)(tD_t)^v\}$. These differential operators are said to be *Fuchsian* with respect to the t-variable. Then $F^{-1} = tF^0$, and so on, while $F^1 = F^0 + D_t F^0$, and so on.

The associated graded ring $\oplus F^k/F^{k-1}$ is not commutative. In fact, one discovers that it is isomorphic to the Weyl algebra itself, namely, to $A_n(\mathbb{C}) \otimes_{\mathbb{C}} A_1$ where A_n is a copy of the Weyl algebra in the x-variables and A_1 is $A_1(t, D_t)$ filtered by $\text{ord}(t) = -1$ and $\text{ord}(D_t) = +1$.

The filtration *does not satisfy the closure condition.* For example, $t^2 D_t \in F^{-1}$ and yet $1 + t^2 D_t$ is not invertible in the ring A_{n+1}. However, the filtration satisfies the comparison condition. This is easy to verify. Using this we take some A_{n+1}-module M and choose a good filtration with respect to the present filtration $\{F^v\}$ and get $\text{gr}_\Gamma(M)$ as a new finitely generated A_{n+1}-module. Now $j(\text{gr}_\Gamma(M)) \geq j(M)$ holds by Theorem 4.4 in Part I. In particular, we see that if M is holonomic then $\text{gr}_\Gamma(M)$ is either the zero module or holonomic.

Example. Consider some $P(x) \in \mathbb{C}[x]$ and take the polynomial $t - P(x)$ in $\mathbb{C}[x,t]$ to get $M = \mathbb{C}[x, t, (t - P(x))^{-1}]$ as a left A_{n+1}-module. Here it is very easy to verify that M is a holonomic A_{n+1}-module. We leave it as a small exercise. For example, a cyclic generator is $1/(t - P(x))$ and if we call it u and put $\Gamma_v = F^v u$ we have a good filtration on M.

2.2. *Formula.* Now $\text{gr}_\Gamma(M) = \oplus \Gamma_v/\Gamma_{v-1}$ is a holonomic A_{n+1}-module. In particular, Γ_0/Γ_{-1} is a module over F_0/F_{-1} and this ring can be identified with $A_n[\nabla]$ where $\nabla = tD_t$ and A_n is the Weyl algebra in the x-variables, i.e., use Formula 1.2 to get this.

Now Γ_0/Γ_{-1} is easy to find. It is nothing but $\mathbb{C}[x, P^{-1}]$ and at this stage we can prove that it is a holonomic A_n-module using the fact that $\text{gr}_\Gamma(M)$ is a holonomic A_{n+1}-module. Namely, the point is that there is a left A_{n+1}-linear mapping θ on $\text{gr}_\Gamma(M)$ which comes from the construction, it is defined by the multiplication with ∇ on Γ_0/Γ_{-1}, and it is in general induced by the multiplication with ∇_{-k} on Γ_k/Γ_{k-1} for all integers k.

Now any A_{n+1}-linear mapping on a holonomic module satisfies an algebraic equation—this is rather easy to prove directly using the fact that holonomic A_n-modules have finite length. See for example [Björk: Chapter 1] for an elementary proof.

In particular, one finds a polynomial $b(\theta)$ so that $b(\theta)\,\mathrm{gr}_\Gamma(M) = 0$. In particular, $b(\nabla)(\Gamma_0/\Gamma_{-1}) = 0$ and then one easily concludes that the A_n-module Γ_0/Γ_{-1} is holonomic. So by Theorem 4.4 in Part I we can obtain a new proof of the holonomicity of $\mathbb{C}[x, P^{-1}]$. At the same time, the minimal polynomial $b(\nabla)$ which annihilates Γ_0/Γ_{-1} is very interesting. It is related to the *Bernstein–Sato polynomial* which enters in functional equations of the form $b(s)P^s = Q(x, D_x, s)P^{s+1}$. See again [4: Introduction] or [1] for this.

2.3. *Remark.* The proof above was a bit sketchy. Time prevents us from giving more details. We mention only that the use of the nondiscrete filtration on A_{n+1} is important to analyze certain geometric constructions which appear in the passage to *vanishing cycles*. See, for example, Malgrange's work [25]. Besides, using similar techniques, we can generalize the preceding material about A_n to the ring of differential operators with analytic coefficients and even to the stalks of the sheaf of micro-local differential operators. So the material in §4 of Part I has extensive applications.

2.4. *Regular holonomic modules.* Among the holonomic A_n-modules one finds a distinguished class which, in addition, are regular. The regularity is defined by geometric constructions and includes conditions at infinity. See, for example, the survey by Katz in [20] for a discussion about regularity. In the lectures by I. N. Bernstein in [2] the reader will find a more detailed study about holonomic A_n-modules with so called "regular singularities" and this class is used to get the *Riemann–Hilbert correspondence* which is established in Bernstein's lectures [20]. We remark that a corresponding theory exists in the analytic case. It requires much more work. See, for example, [19] where the algebraic analysis about holonomic $\mathcal{E}$-modules is pushed very far.

3. Further results. In §1 and §2 we have discussed results about modules over the Weyl algebra $A_n(\mathbb{C})$. In particular, we defined holonomic modules which arise as in §1.24 since $A_n(\mathbb{C})$ is a Noetherian Gorenstein ring. We may also try to use *Krull dimensions* to measure the size of finitely generated $A_n(\mathbb{C})$-modules. We refer to [12] for the definition of Kr.dim(M) when M is a finitely generated A_n-module. Using a counting with multiplicities and the inequality $0 \le j(M) \le n$, one easily verifies that Kr.dim$(M) \le n - j(M)$. In particular, holonomic A_n-modules have Krull dimension zero. This was already observed in the final part of Part I. A recent example due to J. T. Stafford shows that the inequality is strict. In fact, for any $n \ge 2$, one can find a single element P in A_n such that the cyclic A_n-module A_n/A_nP is simple. In particular, Kr.dim$(A_n/A_nP) = 0$ and the j-number is 1.

For example, with $n = 2$, Stafford shows that we can take P as $x_1 + y_1x_2y_2 + x_2 + y_2$.

So the study of Krull dimensions is independent of the methods which we have used in Part I where $j(M)$ was used to measure the size of M when R is a Noetherian Gorenstein ring. If we consider instead of A_n a finite-dimensional Lie algebra J and take its enveloping algebra $U(J)$, then similar phenomena occur. For example, in [31] it is proved that the Krull dimension of $U(\mathrm{sl}(2, \mathbb{C}))$ is 2 while $\mathrm{sl}(2, \mathbb{C})$ is 3-dimensional.

Let us finally mention some references where various questions are treated which are related to the material in these notes: [6, 14, 23].

REFERENCES

[1] Bernstein, I. N., *The analytic continuation of generalized functions with respect to a parameter*, Funz. Analysis Akademia Nauk CCCR **6** (4) (1972), 26–40.

[2] Bernstein, I. N., *Lectures on $\mathcal{D}$-modules* (Preprint to be published in conference at Luminy, July 1983).

[3] Björk, J.-E., *The global homological dimension of some algebras of differential operators*, Inventiones Math. **17** (1972), 67–78.

[4] Björk, J.-E., *Rings of differential operators*, North-Holland Math. Library, Vol. **21** (1979).

[5] Björk, J.-E., *On characteristic varieties*, Journées complexes, Nancy 1982, Revue de l'institut Elie Cartan **8** (1983).

[6] Borho, W. and Brylinsky, J.-L., *Differential operators on homogeneous spaces I: Irreducibility of the associated variety for annihilators of induced modules*, Inventiones Math. **69** (1982), 437–476.

[7] Brylinksi, J.-L., *Modules holonomes à singularités régulières et filtration de Hodge II*, Coll. Analyse et topologie sur les espaces singuliers, Astérisque **101–102** (1983).

[8] Cohn, P. M., *Skew field constructions*, London Math. Soc. Lecture Notes #27.

[9] Fossum, R. M., Griffith, P. A., and Reiten, I., *Trivial extensions of abelian categories, homological algebra of trivial extensions of abelian categories with applications to ring theory*, Springer Lecture Notes **456** (1975).

[10] Gabber, O., *The integrability of characteristic varieties*, Amer. J. Math. **103** (1981), 445–468.

[11] Gabber, O., *Equi-dimensionality of characteristic varieties*, Lectures Paris, 1981 (revised by Levasseur).

[12] Gabriel P., and Rentschler, R., *Sur la dimension des anneaux et ensembles ordonnés*, C. R. Acad. Sci. Paris Sér. A **265** (1967), 712–715.

[13] Godement, R., *Topologie algébrique et théorie des faisceaux*, Actual. Sci. et Industr. **1252**, Paris, Hermann, 1958.

[14] Goodearl, K. R., *Global dimensions of differential operators II*, Trans. Amer. Math. Soc. **209** (1975), 65–85.

[15] Hironaka, H., *Resolution of singularities of an algebraic variety over a field of characteristic zero, I and II*, Ann. of Math. **79** (1964), 109–203, 205–326.

[16] Kashiwara, M., *Algebraic study of systems of partial differential equations*, Master's thesis, University of Tokyo (1971). (In Japanese)

[17] Kashiwara, M., *Systems of microdifferential equations*, Birkhäuser, Progress in Mathematics **34** (1983).

[18] Kashiwara, M. and Kawai, T., *Microlocal analysis*, RIMS Kyoto University, February, 1983 (preprint).

[19] Kashiwara, M. and Kawai, T., *Second-microlocalization and asymptotic expansions*, Lecture Notes in Physics, Springer-Verlag **126**.

[20] Katz, M., *The regularity theorem in algebraic geometry*, Actes Congress Inter. Math. Nice, Vol. 1 (1970), 437–443.

[21] Lê Dũng Tráng and Mebkhout, Z., *Variétés caractéristiques et variétés polaires*, C. R. Acad. Sci. Paris Sér. I Math. **296** (1983), 129–132.

[22] Lê Dũng Tráng and Mebkhout, Z., *Introduction to linear differential systems*, Proc. Sympos. Pure Math. **40**, Amer. Math. Soc., Providence, R. I. (1983).

[23] Levasseur, T., Thèse d'état, Paris, 1984 (?).

[24] Malgrange, B., *L'involutivité des caractéristiques des systèmes différentiels et microdifférentiels*, Sém. Bourbaki **522** (1978), 277–289.

[25] Malgrange, B., *Polynômes de Bernstein–Sato et cohomologie évanescente*, Astérisque **101–102** (1983), 243–267.

[26] Roos, J.-E., *Détermination de la dimension homologique globale des algèbres de Weyl*, C. R. Acad. Sci. Paris Sér. A-B **274** (1974), A23–A26.

[27] Roos, J.-E., *Compléments à l'étude des quotients primitifs des algèbres enveloppantes des algèbres de Lie semi-simples*, C. R. Acad. Sci. Paris Sér. A-B **276** (1973), A447–A450.

[28] Sato, M., Kashiwara, M., and Kawai, T., *Microfunctions and pseudo-differential equations*, Springer Lecture Notes **287** (1973).

[29] Sato, M. and Kashiwara, M., *The determinant of matrices of pseudo-differential operators*, Proc. Japan Acad. **51** (1975), 17–19.

[30] Sato, M., Kashiwara, M., Kimura, T., and Oshima, T., *Microlocal analysis of prehomogeneous vector spaces*, Inventiones Math. **62** (1980), 117–179.

[31] Smith, S. P., *Krull dimension of the enveloping algebra of* $\mathrm{sl}(2,\mathbb{C})$, J. Algebra **71** (1981), 189–194.

[32] Stafford, J. T., *Nonholonomic modules over Weyl algebras and enveloping algebras*, Invent. Math. **79** (1985), 619–638.

Noetherian Group Rings: An Exercise in Creating Folklore and Intuition

DANIEL R. FARKAS

This is Not a Survey

My intention is two-fold. I wish to expose some underlying Noetherian themes in the drama of group rings. The stars of this production are Philip Hall and Jim Roseblade, with a crowd-pleasing cameo appearance by George Bergman. In addition, I would like to give some indication why group theorists are interested in these techniques. For them, the relevant device might be termed "Noether ex machina".

To understand the tone of these notes, the reader should realize that the talk on which they are based was delivered during the last day of an intense five day conference. Few attentions can span such a program. I have highlighted material by condensing, lying, conjecturing, punning, and taking a generally irreverent attitude. I hope I kept some people awake. There is no doubt in my mind that I have divided the practitioners of group rings into two camps: those who will be insulted because they were not mentioned and those who will be insulted because I butchered their best work when I discussed it.

This article is not the place to look for a comprehensive tour of the subject. I have not always chosen the deepest or prettiest theorems to discuss. Instead, I have looked for vehicles that develop intuition. If my idiosyncratic approach succeeds, the reader will have an idea of how to think globally about the subject, without really having any idea how to prove any local result. To paraphrase Don Passman [**16**], I hadn't meant for this report to be so eccentric. But once I threw out all of the rules, I had no choice.

I will be discussing the group ring of a polycyclic-by-finite group. To review, G is said to be polycyclic-by-finite if it has a finite series

$$1 = G_0 \subset G_1 \subset \cdots \subset G_m = G$$

with G_{j-1} normal in G_j and G_j/G_{j-1} either finite or infinite cyclic. It's possible to push all of the finite layers to the very top factor and assume that G_j/G_{j-1} is infinite cyclic for $j < m$. The group algebra $k[G_{m-1}]$ then looks much like an iterated twisted polynomial ring.

As a consequence, there are many propositions about the group algebra of a polycyclic-by-finite group whose proofs are twisted versions of polynomial ring arguments. I will generally ignore such considerations. This is not to denigrate the technique; its execution frequently requires cleverness and elucidates the basic ring-theoretic nature of many theorems. However, I want to stress those methods which are uniquely group algebra-theoretic.

Fortunately for group theorists (and unfortunately for this presentation) polycyclic groups can be quite complicated. I have decided to adopt the fiction that there are only two polycyclic-by-finite groups, feeling that we have to settle for the flavor when we can't afford the whole meal. Here are the groups.

H is the "discrete Heisenberg group." It is generated by x, y, and z subject to the relations that z be central and that $[x, y] = z$. Concretely, it can be represented as the group of 3×3 unitriangular matrices with integer entries. Obviously H has a normal series

$$1 \subset \langle z \rangle \subset \langle z, x \rangle \subset \langle z, x, y \rangle = H$$

with infinite cyclic factors. We set $\mathbb{A} = \langle z \rangle$, the center of H. Notice that the group is nilpotent. (Every finitely generated nilpotent group is polycyclic.)

B is the simplest interesting nonnilpotent polycyclic group. (It will be used to illustrate the ubiquity of Bergman's Theorem.) It is a semidirect product $(\mathbb{Z} \oplus \mathbb{Z}) \rtimes \langle w \rangle$ where w acts on $\mathbb{Z} \oplus \mathbb{Z}$ like the matrix $\left(\begin{smallmatrix} 1 & 2 \\ 1 & 1 \end{smallmatrix}\right)$. We will find it useful to record that the eigenvalues of $\left(\begin{smallmatrix} 1 & 2 \\ 1 & 1 \end{smallmatrix}\right)$ are the real irrational numbers $1 \pm \sqrt{2}$. When we speak of B, the notation $\mathbb{A}$ will designate the copy of $\mathbb{Z} \oplus \mathbb{Z}$ which is the normal abelian subgroup "at the bottom."

The HilbARt Basis Theorem

There would be no story to tell without Philip Hall's twisted version of the Basis Theorem—if S is a Noetherian ring (pick a side) and G is a polycyclic-by-finite group then $S[G]$ is Noetherian [**6**]. While this proposition is now regarded as just a hand-wave away from the commutative Basis Theorem, the same argument (when applied to a polynomial-like overring concocted from a given ideal) gives rise to the powerful Artin–Rees property. Though connections to group theory had already been recognized by Hall himself, Jategaonkar reawakened interest in the property by exploiting it to complete the solution to one of Hall's long-standing problems [**10**].

Recall that a two-sided ideal I has the weak Artin–Rees property in the ring R (or that I "has AR", for short) provided that for each right ideal E in R there is an integer $n \geq 1$ such that $E \cap I^n \subseteq EI$. We frequently refer to an equivalent formulation for modules: if M is a finitely generated R-module and U is an essential submodule of M such that $UI = 0$ then there is a positive integer n with $MI^n = 0$. When the Artin–Rees property is present, a Krull intersection theorem is not far behind. Roughly speaking, if I has AR in a sufficiently Noetherian prime ring then $\bigcap_{d=1}^{\infty} I^d = 0$.

Ideals in a Noetherian ring have AR if they are generated in a sufficiently commutative manner. The workhorse in the theory of group rings is the following proposition of Jategaonkar [**10**] and Roseblade [**20**].

THEOREM. *Let R be a ring containing a polycyclic-by-finite group of units G and a (right) Noetherian subring S. Assume that G normalizes S and that R is generated by S and G. If I is a polycentral ideal of S that is normalized by G then $RI = IR$ has* AR.

Here I is polycentral if it is generated by elements $0 = x_0, x_1, \ldots, x_n$ such that x_i is central modulo the ideal generated by $x_0, x_1, \ldots, x_{i-1}$. Where does one find polycentral ideals? Certainly ideals in the group algebra of a finitely generated abelian group are polycentral. It is not difficult to imagine that the same is true for all ideals in the group algebra of a finitely generated nilpotent group. (More about this later.) As a consequence, if G is polycyclic-by-finite then an ideal, blown up from an invariant ideal in the group algebra of a normal nilpotent subgroup, has AR. In particular, every ideal in $k[H]$ has the weak Artin–Rees property.

What happens for ideals of $k[B]$? Let's look at the augmentation ideal I in the case that $\operatorname{char} k \neq 2$. Choose an arbitrary $a = (u, v) \in \mathsf{A}$ and set $b = (2v, u)$.

$$wbw^{-1}b^{-1} - 1 = [(w-1)(b-1) - (b-1)(w-1)]w^{-1}b^{-1} \in I^2.$$

But $(wbw^{-1})b^{-1} = \begin{pmatrix} 0 & 2 \\ 1 & 0 \end{pmatrix}\binom{2v}{u} = \binom{2u}{2v} = a^2$. Hence $a^2 - 1 \in I^2$. Thus

$$a - 1 = \tfrac{1}{2}((a^2-1) - (a-1)^2) \in I^2.$$

We see now that the argument can be repeated to prove that $a - 1 \in \bigcap I^n$. This is incompatible with the Krull intersection theorem one expects if I satisfied AR.

Paradise is almost regained for fields of positive characteristic. Roseblade [**20**] proves that if G is a polycyclic-by-finite group and the characteristic of k is a prime p then there is a characteristic subgroup $G_{(p)}$ of finite index in G such that every ideal of $k[G_{(p)}]$ has AR.

We can see how this might happen by re-examining $k[B]$ when k is a field of positive characteristic p. The canonical image of $\begin{pmatrix} 1 & 2 \\ 1 & 1 \end{pmatrix}$ in $\mathrm{GL}(2, \mathbb{Z}/(p))$ is invertible and so has finite order m. In terms of the group B, conjugation by w^m sends an element a in A to ab^p for some $b \in \mathsf{A}$. We shave B slightly to obtain the subgroup B' of finite index, which is generated by A and w^m. This time let I be the augmentation ideal of $k[B']$ and assume that U is an essential submodule of M with $UI = 0$. The AR property *does* hold for the augmentation ideal of $k[\mathsf{A}]$, which sits inside I: there is a positive integer, which we might as well write in the form p^t, such that $M(a-1)^{p^t} = 0$ for all $a \in \mathsf{A}$. Of course, over a field of characteristic p we have $(a-1)^{p^t} = a^{p^t} - 1$. If we let A_j denote the characteristic subgroup of A consisting of all p^j-powers, then we are saying that members of A_t act like 1 on M. Surely we can replace B' by B'/A_t. Look at the action of w^m on the factors of the characteristic series

$$1 = \frac{\mathsf{A}_t}{\mathsf{A}_t} \subseteq \frac{\mathsf{A}_{t-1}}{\mathsf{A}_t} \subseteq \cdots \subseteq \frac{\mathsf{A}_0}{\mathsf{A}_t}.$$

Each factor is a $\mathbb{Z}/(p)$-vector space on which w^m acts like the identity matrix. We have produced a central series, thereby proving that $B'/\mathbb{A}_t$ is nilpotent. In particular, AR holds for all ideals of its group algebra. Hence $UI = 0$ implies $MI^n = 0$ for an appropriate n.

I'd like to give a very pretty application of the AR property due to Ken Brown, as found in Donkin's paper. It illustrates the basic strategy of reducing problems to the case of polynomial identity algebras.

THEOREM ([4]). *Suppose k is an arbitrary field and G is a polycyclic-by-finite group. The injective hull of a finite-dimensional $k[G]$-module is locally finite-dimensional.*

Proof. We quickly reduce to proving that if W is a finitely generated essential extension of a direct sum $V = V_1 \oplus \cdots \oplus V_n$ of finite-dimensional simple modules then W is finite-dimensional. I'll concentrate on the essence of the argument by restricting attention to our two canonical groups and by assuming that V is just the module k with the trivial action of G. The crucial fact is that $\mathbb{A}$ and $G/\mathbb{A}$ are abelian. (A fundamental theorem of Mal′cev says that every polycyclic group is nilpotent-by-(abelian-by-finite).)

Blow up the augmentation ideal of $k[\mathbb{A}]$ to an ideal I of $k[G]$. We have seen that I has AR. Also, $VI = 0$ since $v \cdot a = v$ for $a \in \mathbb{A}$ is equivalent to $v \cdot (a - 1) = 0$. Since V is essential in W there is an integer n such that $WI^n = 0$. If $R = k[G]/I^n$ then R has a nilpotent ideal (namely I) which, when factored out, yields a commutative algebra (namely $k[G/\mathbb{A}]$). Therefore R is a PI-algebra. By a theorem of Jategaonkar [**9**], W is an Artinian R-module and so has finite composition length. On the other hand, a simple R-module is a simple $k[G/\mathbb{A}]$-module; the ordinary Nullstellensatz implies that such modules are finite-dimensional over k. Hence W is finite-dimensional over k. □

This is just the first step in a much deeper theory which develops an analogue to Matlis' analysis of injective hulls for simple modules over a commutative Noetherian ring. For instance, the injective hull of a finite-dimensional $k[G]$-module is Artinian. For k of positive characteristic this is due to Jategaonkar [**11**] and Musson [**14**]. The gist is that Roseblade supplies an ideal of finite codimension, in the annihilator of the module, which has strong AR properties. This allows one to complete the group ring and set up a duality between submodules of the injective hull and ideals in the Noetherian completion. The program for k of characteristic zero is less obvious. Donkin [**4**] handles this case by analyzing a sort of algebraic group (actually a nearly affine Hopf algebra) that is assigned to each polycyclic group; there is a categorical equivalence between locally finite-dimensional $k[G]$-modules and co-modules over the Hopf algebra. (Interestingly enough, the required affineness is checked via a twisted polynomial argument.) Apparently, the advantage of using the algebraic group is that there is a splitting into unipotent and reductive pieces.

I'll close this section with the intriguing remark that one can sometimes buy Noetherian at the cost of AR. I refer to a theorem of P. Smith [**24**] stating that if k is a field or the integers and if the augmentation ideal of $k[G]$ has AR then localization of the group ring at those elements with augmentation 1 (after factoring out their "torsion") is possible and yields a right and left Noetherian ring! Snider has exploited this idea to prove the zero divisor conjecture for some solvable groups which are not polycyclic [**25**].

Nullstellensatz and Friends

The other fundamental Noetherian theorem is Hilbert's Nullstellensatz. The most successful approach to noncommutative analogues has been via generic flatness. I would argue that this point of view was first taken by Philip Hall [**7**].

TWISTED GENERIC FLATNESS (Hall, as refined by Roseblade). *Let R be a ring containing a polycyclic-by-finite group of units G and a commutative Noetherian domain S. Assume that G normalizes S and that R is generated by S and G. If M is a finitely generated R-module then there is a nonzero element f in S such that M_{f*G} is a free S_{f*G}-module.* (We are localizing at the set $f * G$ of all products $f^{x(1)} f^{x(2)} \dots f^{x(n)}$ where $x(i) \in G$.)

The argument is of the twisted polynomial type. Notice that if we actually have a group ring $R = S[G]$ then G centralizes all elements of S and we get ordinary generic flatness. This has the usual Nullstellensatz consequences, e.g., if S is a Jacobson ring, so is $S[G]$. (Recall that a Jacobson ring is a (right) Noetherian ring whose prime images are semiprimitive.)

COROLLARY (Roseblade [19]). *Let S be a commutative Jacobson ring. A simple $S[G]$-module M is killed by some maximal ideal of S.*

Proof. Without loss of generality $\operatorname{ann}_S M = 0$. We wish to show that S is a field. Now M is a torsion free S-module, so S must be a domain. Since S is a Jacobson ring, it is semiprimitive: for example, given a flattener f of M there is some maximal ideal L of S which does not contain it. The localization M_f is a free S_f-module, whence $M_f L \neq M_f$. Therefore $ML \neq M$. By simplicity, $ML = 0$. This contradicts torsion freeness unless $L = 0$. □

In particular, every simple $\mathbb{Z}[G]$-module is really a $\mathbb{Z}/(p)[G]$-module for some prime p.

I present an elegant application to group theory of the ring theory we've discussed so far.

THEOREM (Robinson [18], Wehrfritz [26]). *Let G be a finitely generated solvable group. If all finite homomorphic images of G are nilpotent then G is nilpotent.*

Proof. The proof is by "induction". Roughly speaking, Zorn's Lemma lets us assume that all proper homomorphic images of G are nilpotent. The finite homomorphic image hypothesis can then be exploited to show that we may suppose G has no finite normal subgroups.

Look at the bottom of the derived series of G. There is a normal abelian subgroup A of G which is finitely generated as a G-operator group and either

A is a torsion free abelian group, or
A is an infinite elementary abelian p-group.

In other words, A is either a $\mathbb{Z}$-torsion free finitely generated $\mathbb{Z}[G]$-module or a finitely generated $\mathbb{Z}/(p)[G]$-module. (Warning: we will hop around between the multiplicative view of A as a normal subgroup of G and the additive view of A as a $\mathbb{Z}[G]$-module under the action of conjugation, without further comment.)

By induction, G acts like a finitely generated nilpotent group $\overline{G}$ of automorphisms on A. (Hence A is a $\mathbb{Z}[\overline{G}]$- or $\mathbb{Z}/(p)[\overline{G}]$-module.)

The torsion free case

There is an integer f such that the localization A_f is a free $\mathbb{Z}_f$-module. If q is a prime not dividing f then

$$A_f/A_f \cdot q \simeq A/A \cdot q.$$

But G/A^q is a finitely generated nilpotent group, so A/A^q is finite. (This is the "Noetherian nature" of polycyclic groups.) Therefore A_f has finite rank ρ, so $A/A \cdot q$ cannot have a chain of more than $\rho + 1$ subgroups. Using the nilpotence of G/A^q, this implies

$$\underbrace{[A, G, G, \ldots, G]}_{\leftarrow \rho \rightarrow} \subseteq A^q.$$

The choice of ρ is the same for each q; since G/A is nilpotent we see that there is an m with the mth term of the descending central series $\gamma_m(G) \subseteq \bigcap_{q/f} A^q$. (First commute G with itself down to A and then do it ρ more times.) But this intersection is zero in the free module A_f. (After all, how many primes can you be divisible by?) Hence $\gamma_m(G)$ is a subgroup of A consisting of elements each killed by some power of f. This cannot happen if A is torsion free.

The elementary abelian case

Since A is a finitely generated $\mathbb{Z}/(p)[\overline{G}]$-module which is not finite, $\overline{G}$ cannot be finite. Let z be an element of infinite order in the center of $\overline{G}$. (This element exists as an example of the rule of thumb: as the center goes, so goes the nilpotent group.) Set $S = \mathbb{Z}/(p)[t, t^{-1}]$ and have t act on A like z. Then A is a finitely generated $S[G]$-module. The argument we just made adapts to this case as well, once it is established that A is a torsion free S-module.

First one proves that if the collection T of torsion elements is nontrivial then some power of z, say z^n, centralizes T. Secondly, T is an essential $\mathbb{Z}/(p)[\overline{G}]$-submodule of A. Indeed, if A_0 is a submodule of A with $A_0 \cap T = 0$ then A_0 also functions as a normal subgroup of G and $G \hookrightarrow G/A_0 \times G/T$. By induction, the cartesian product is nilpotent, whence G is too. (Of course, we are implicitly assuming the contrary.)

Let I be the ideal of $\mathbb{Z}/(p)[\overline{G}]$ generated by the central element $z^n - 1$; notice that I has AR! Hence $TI = 0$ implies $AI^d = 0$ for some d. Since we can take $d = p^e$ for some $e \geq 0$ we discover that $(z^n - 1)^{p^e} = z^{np^e} - 1$ kills A. In other words, some power of z acts on A like it has finite order. This is a contradiction. □

Dan Segal wrote a marvelous paper (based on reductions of Lennox) analyzing centrality properties of solvable groups with chain conditions [**23**]. His approach is a considerable deepening of the connections among finitely generated modules over polycyclic group rings, AR-like conditions, and central series that we saw in the Robinson–Wehrfritz Theorem. In ring-theoretic terms, Segal applies the machinery of Krull dimension to prove that every finitely generated module over the integral group ring of a finitely generated nilpotent-by-finite group has a finite series whose factor modules have radical zero. (That is, the intersection of their maximal submodules is zero.) Anyone who doubts that ring theorists ought to be talking to group theorists should be required to read this work.

There is a powerful and deep variation of the Nullstellensatz unique to group algebras. It was conjectured by Hall (who verified it for nilpotent groups) and later proved by Roseblade.

THEOREM (Roseblade [19]). *Let k be an absolutely algebraic field, (that is, an algebraic extension of a finite field). If G is a polycyclic-by-finite group then simple $k[G]$-modules are finite-dimensional over k.*

Combined with the corollary to generic flatness, this theorem asserts that simple $\mathbb{Z}[G]$-modules are actually finite.

Proof. Let's assume that k is a finite field. As usual, G is one of our two canonical groups. We first argue that the given simple module M has torsion as a $k[\mathbb{A}]$-module.

What happens when M is torsion free? Find a nonzero flattener f for M, as a $k[\mathbb{A}]$-module. As in the corollary to generic flatness, one would expect that if L is a maximal ideal of $k[\mathbb{A}]$ which misses all conjugates of f then $M_{f*G}L \neq M_{f*G}$. For such L we have $ML \neq M$. On the other hand, it's possible to exploit the cofiniteness of maximal ideals in $k[\mathbb{A}]$ together with the assumption that M is torsion free to conclude that $ML = M$ for all maximal ideals L. (More on the advantages of using finite coefficient fields can be found in the appendix.) We conclude that every maximal ideal of $k[\mathbb{A}]$ will contain a conjugate of f. Let's see what's wrong with this in our two cases.

H

Since the action of G on $\mathbb{A}$ is central here, f lies in the intersection of all the maximal ideals of $k[\mathbb{A}]$. That is, f is zero—a contradiction.

B

We play the same game by just looking at the B-invariant maximal ideals of $k[\mathbb{A}]$, those invariant under the generator w. Are there "enough" of these? A maximal ideal is realized as the kernel of a map into a finite extension field of k. By carefully estimating the number of invariant ones as the size of the extension field grows, we find there are infinitely many. (This counting argument is redolent of the zeta function for varieties.) Their intersection is a B-invariant ideal of $k[\mathbb{A}]$ which does not have finite codimension. Question: are there any, other than zero? If not, we have the same contradiction as before since f lies in the intersection.

Don't forget the question.

We complete the argument that M is finite-dimensional over k.

H

Since M is a simple $k[G]$-module and $k[\mathbb{A}]$ is centralized by G, we see that anything in $k[\mathbb{A}]$ that kills a nonzero element of M kills all of M. Hence $I = \operatorname{ann}_{k[\mathbb{A}]} M$ is not zero. But $k[\mathbb{A}]$ is just the group ring of an infinite cyclic group: thus I has finite codimension in $k[\mathbb{A}]$. It then swallows most of $\mathbb{A}$, so it is possible to rig an induction argument to complete the proof. Alternatively, M can now be regarded as a simple module over the affine PI-algebra $k[G]/Ik[G]$. Primitive images of this ring are finite-dimensional over their centers. The Artin–Tate argument tells us that each of these centers is finitely generated over k. Finally, an application of the Nullstellensatz yields each primitive image finite-dimensional over k, whence M is finite-dimensional.

B

Again, we essentially find a nonzero annihilator ideal I which is B-invariant. But this time we cannot automatically conclude that $k[\mathbb{A}]/I$ is finite-dimensional because $k[\mathbb{A}]$ is like a polynomial ring in two variables. So we are stuck with the same question as before: are there B-invariant ideals of $k[\mathbb{A}]$ of infinite codimension other than zero? □

A theorem has forced itself upon us. (I believe that Zalesskii noticed its inevitability first.) The hypothesis should be about an abelian group which is "small" with respect to a group of operators. The conclusion should limit the possible invariant ideals of the abelian group algebra. Let A be a finitely generated torsion free abelian group and suppose that G is a group which acts on A. Then A is said to be a "plinth" for G if the rational vector space $\mathbb{Q} \otimes_{\mathbb{Z}} A$ is irreducible under the action of every subgroup with finite index in G. (The terminology is, of course, classical.) An equivalent formulation is that the only subgroups of A with finitely many G-conjugates are those with finite index in A.

It is obvious that $\mathbb{A}$ is a plinth for the action (by conjugation) of H; you can't fit much into an infinite cyclic group. To see that $\mathbb{A}$ is a plinth for the action of B, notice that a rationalized one-dimensional subspace invariant under a subgroup of finite index in B corresponds to an eigenvalue of ± 1 for some power of w. But the eigenvalues of w are real irrational numbers. They cannot be roots of unity.

We arrive at the basic theorem, due to George Bergman.

THEOREM ([1]). *Suppose that A is a plinth for the action of an arbitrary group G. Then all nonzero G-invariant ideals of $k[A]$ have finite codimension over k. (That is,* $\dim_k k[A]/I < \infty$.)

Appendix I

I would like to step back for a moment and muse about the role of absolutely algebraic fields in Roseblade's Theorem. Initially, this seems foolish. Firstly, there are always infinite-dimensional simple modules over other fields unless the group in question is abelian-by-finite—the PI case. (This observation of Hall has been generalized by a gang of people.) Secondly, the proof depends on a counting argument and other uses of finite fields that I did not make explicit. However, Harper has observed that large portions of Roseblade's argument hold for arbitrary fields via a "Brauer lifting argument" [**8**]. Musson formulates the lifted theorem in the following way.

THEOREM ([15]). *Assume k is any field and A is a free abelian normal subgroup of the polycyclic-by-finite group G. Suppose, further, that A is a plinth for the action of G by conjugation. If M is a simple $k[G]$-module then the restriction $M|_{k[A]}$ is a locally finite-dimensional $k[A]$-module.*

(By the way, every infinite polycyclic group has a subgroup of finite index with such a plinth.)

The property of an absolutely algebraic field F which underlies Roseblade's Nullstellensatz is an elementary application of the ordinary Nullstellensatz: if B is a finitely generated abelian group and if the image of B in $F[B]/I$ is infinite then the ideal I has infinite codimension in $F[B]$. (This should ring a bell.) To see that this makes the statement of Roseblade's Theorem a special case of Musson's, note that we may assume that the image of G in $F[G]/\operatorname{ann} M$ is faithful. Now we are positing that some element of M is killed by an ideal of $F[A]$ with finite codimension. By the crucial property of F, there is some positive integer ν and some nonzero $m \in M$ such that

$$m(a^\nu - 1) = 0 \text{ for all } a \in A.$$

But the subgroup $\{a^\nu \mid a \in A\}$ is characteristic,

$$0 = m(a^\nu - 1)F[G] = mF[G](a^\nu - 1) = M(a^\nu - 1).$$

We conclude that $a^\nu = 1$ for all $a \in A$. Since plinths are torsion free, it must be that G is a *finite* group.

The Primes of Mr. Jim Roseblade

We turn to the premier result in the area, Roseblade's Prime Controller.

THEOREM ([21]). i. *If G is a polycyclic-by-finite orbitally sound group and P is a faithful prime ideal of $k[G]$ then $P = (P \cap k[\Delta]) \cdot k[G]$. (Here Δ denotes the finite conjugate subgroup of G.)*

ii. *Every polycyclic-by-finite group contains a subgroup of finite index which is orbitally sound.*

I think it may be more valuable to describe some of the characteristics of orbitally sound groups and develop intuition as we go along, rather than ever define the term. For instance, homomorphic images and finite index subgroups of orbitally sound groups are orbitally sound. A finitely generated nilpotent group is orbitally sound. Most of the time the Prime Controller is used, one descends to an orbitally sound subgroup of finite index with additional special properties. This frequently corresponds to reducing to the connected component in the study of linear groups. That point of view is more than a meta-theorem. Every polycyclic-by-finite group has an imbedding in $\mathrm{GL}(n, \mathbb{Z})$. Wehrfritz observed that its connected component, with respect to the relative Zariski topology, is always orbitally sound. I'll offer a bit of intuition-building. If G acts faithfully on $\mathbb{Z}^n$ then it imbeds in $\mathrm{GL}(n, \mathbb{Z})$; it is easy to see that $\mathbb{Z}^n$ is a plinth if and only if the connected component of G acts irreducibly on the underlying vector space $\mathbb{Q}^n$. Understanding plinths amounts to understanding "irreducible" orbitally sound groups.

An ideal P in $k[G]$ is faithful if the *group* homomorphism sending G to its image in $k[G]/P$ is faithful. When the ideal P is not faithful, $k[G]/P$ factors through $k[\overline{G}]$ where $\overline{G}$ is the appropriate homomorphic image of G; P is carried to an ideal of $k[\overline{G}]$ which is now faithful. Since homomorphic images of polycyclic groups are polycyclic, we see that it is always possible to restrict attention to faithful ideals.

The finite conjugate subgroup $\Delta(G)$ is comprised of those elements in G which have finitely many conjugates. Let's identify this subgroup for our examples. Any element g in H which is outside the center has an x-conjugate or y-conjugate different from g: say $xgx^{-1} = z^l g$ where $l \neq 0$. Then $x^m g x^{-m} = z^{ml} g$, thereby producing infinitely many conjugates. The upshot is that $\Delta(H)$ is the center of H. Next, we look for $\Delta(B)$. First notice that $\Delta(B) \cap \mathbb{A} = \mathbf{1}$. Otherwise some power of w centralizes a nonidentity element of $\mathbb{A}$; we have already seen that this constrains the "eigenvalues" of w in an impossible way. Hence either $\Delta(B) = 1$ or $\Delta(B)$ is infinite cyclic. In the latter case, the square of every element of B must centralize $\Delta(B)$. The generator of $\Delta(B)$ then acts on $\mathbb{A}$ like a nonzero power of w which has a fixed point. We reach the same contradiction as before, unless $\Delta(B) = 1$.

In general, the center of $\Delta(G)$ has finite index in $\Delta(G)$. One consequence is that $k[\Delta(G)]$ is a PI-algebra. Actually, most applications of Roseblade's Prime Controller reduce to the case that Δ is abelian.

Most of this address will be devoted to the consequences of Roseblade's theorem. However, it should not be compared with the program describing prime ideals in the theory of enveloping algebras. It is not likely to give insight into a global picture of prime ideals; primes here are controlled one at a time. Partly this is due to the restriction to faithful primes. (This is inevitable because polycyclic groups have too many finite images.) More fundamentally, very little is known about the orbit structure of a variety induced from a multiplicative, rather than affine, action. Finally, descending to an orbitally sound subgroup of finite index is fatal to a global analysis. For example, the description of prime ideals in the group algebra of an abelian-by-finite group is at the same preliminary stage as the description of the spectrum of an affine PI-algebra.

Before proceeding to a "proof", I want to emphasize that I am skirting some wonderful and deep ideas due to Roseblade. His paper is a landmark in algebra.

Proof. The first step is due to Zalesskii [**27**]. We show that if G is one of our two groups and P is a faithful prime of $k[G]$ then $P = (P \cap k[\mathbb{A}]) \cdot k[G]$.

H is a nilpotent group (remember that $\mathbb{A}$ is the center and $H/\mathbb{A}$ is abelian), so it is not surprising that $k[H]$ is polycentral: any nonzero ideal in a homomorphic image of $k[H]$ contains a nonzero central element. Now B is not nilpotent; however, it has a normal series $\mathbb{A} \triangleleft B$ such that $\mathbb{A}$ acts trivially on $\mathbb{A}$ and on $B/\mathbb{A}$. Hence $\mathbb{A}$ acts trivially on some proper normal subgroup of every homomorphic image of B. With this information, the proof of centrality can be souped up to prove that every nonzero ideal in a homomorphic image of $k[B]$ contains a nonzero element centralized by $\mathbb{A}$. (This line of ideas is found in Roseblade–Smith [**22**].)

Now suppose that P is a faithful prime of $k[G]$. Let's factor out $(P \cap k[\mathbb{A}]) \cdot k[G]$ and study a possible nonzero element in the image of P which is either centralized by H (in the case $G = H$) or by $\mathbb{A}$ (in the case $G = B$). It ends up good enough to assume that this element is the image of some ax where $a \in k[\mathbb{A}]$ and x is in G. Under the assumption that the image of ax is nonzero, we have $a \notin P \cap k[\mathbb{A}]$. Consider the minimal primes of $k[\mathbb{A}]$ over $P \cap k[\mathbb{A}]$:

$$Q_1, Q_2, \ldots, Q_n.$$

It's not difficult to see that they comprise a single orbit under the action of G. One of these primes, call it Q_*, misses a.

The case $G = H$

By centrality, if $g \in H$ then $ax^g - ax \in (P \cap k[\mathbb{A}]) \cdot k[H]$. Thus $a(x^g x^{-1} - 1) \in P \cap k[\mathbb{A}]$. (You have to notice that $x^g \equiv x (\operatorname{mod} \mathbb{A})$ for all x and g in H.) Back to our list of primes, $\mathbb{A}$ is central in H, so there is really only one Q there and it must coincide with $P \cap k[\mathbb{A}]$. In other words, $P \cap k[\mathbb{A}]$ is a real live prime in a commutative ring:

$$a \notin P \cap k[\mathbb{A}] \text{ and } a(x^g x^{-1} - 1) \in P \cap k[\mathbb{A}] \Rightarrow x^g x^{-1} - 1 \in P.$$

How can $x^g x^{-1}$ be sent to 1 modulo P if it is faithful? We must have $x^g = x$ for each $g \in H$. This means that x lies in the center and, hence, that ax lies inside $k[\mathbb{A}]$ as well as P. We have argued that the image of ax modulo $(P \cap k[\mathbb{A}]) \cdot k[H]$ is zero, a contradiction. Since $\mathbb{A} = \Delta(H)$, the Prime Controller is proved for this case.

The case $G = B$

The argument begins in a similar way except that $P \cap k[\mathbb{A}]$ is not one of the Q_i. So let S_i be the subgroup of $\mathbb{A}$ sent to 1 after factoring out Q_i. Then the S_i are B-conjugate and $S_1 \cap \cdots \cap S_n = 1$ by fidelity. This obviously can't happen in a plinth unless each $S_i = 1$. In particular, $S_* = 1$. Argue as above, with Q_* replacing $P \cap k[\mathbb{A}]$, that $P = (P \cap k[\mathbb{A}]) \cdot k[B]$. Now a typical Q_j has finitely many conjugates, so it is left invariant by a subgroup B' with finite index in B, still containing $\mathbb{A}$. It's easy to see that $\mathbb{A}$ remains a plinth for B' and that Q_j is a faithful prime ideal of $k[B']$. By Bergman's Theorem, either $Q_j = 0$ or $\dim_k k[\mathbb{A}]/Q_j$ is finite.

Let's eliminate the second possibility. Otherwise $k[\mathbb{A}]/Q_j$ is a finite field extension $K|k$ generated by k and the image $i(\mathbb{A})$ in K. The action of B' on $\mathbb{A}$ induces an action of B' on K which fixes k. Elementary Galois theory says that B' acts like a finite group of automorphisms on K, so it acts like a finite group on $i(\mathbb{A})$. But i is faithful! Thus B' is acting like a finite group of automorphisms on $\mathbb{A}$, a contradiction.

We find that $Q_j = 0$. Therefore $P \cap k[\mathbb{A}] = 0$, which implies $P = 0$. This is, of course, the only way to have $P = (P \cap k[1]) \cdot k[B]$. □

A Periodic Problem

A substantial portion of the literature about Noetherian rings is devoted to localization. Even a cursory reading indicates that localizing in noncommutative rings is more delicate and technical than in commutative rings. Rather than illustrate the state of the art by using group rings as an example, I will remind everyone that localization began life as a tool.

Can one localize "enough" in the group algebra of a polycyclic-by-finite group?

THEOREM (cf. Jategaonkar [12]). *Assume that k is either a field or $\mathbb{Z}$ and that G is a polycyclic-by-finite group. If P is a prime ideal of $k[G]$ satisfying (right) AR then P is (right) localizable.*

(In saying that P is right localizable I mean that the ring satisfies the right Ore condition with respect to $\mathcal{C}(P)$.) Pieces of this theorem are scattered throughout the literature. Most of it was proved by Brown, Lenagan, and Stafford [**2**] as a by-product of their investigation of weak-ideal invariance. Recently, Passman showed that faithful completely prime ideals always have AR and so are localizable [**17**].

I will outline Lichtman's use of "enough localizability" in his solution to a Burnside problem. Let G be a polycyclic-by-finite group whose group algebra $k[G]$ is prime and have D denote its classical ring of quotients. Then a group of periodic matrices with entries in D is locally finite [**13**].

To anyone who has ever looked at finitely generated linear groups the strategy is obvious—specialize! Let's carefully review the procedure for a noncommutative Noetherian ring. We have a right and left Noetherian prime ring R and matrices $A_1, \ldots, A_m \in \mathrm{GL}(t, Q(R))$. Since the entries of the matrices lie in the ring of quotients of R, we may rewrite $A_j = a^{-1}B_j$ where $a \in \mathcal{C}(0)$ and $B_j \in \mathrm{GL}(t, R)$. The idea is to find a prime ideal P which "avoids" a and then factor out P. So let's assume that $a \in \mathcal{C}(P)$ as well. It is not difficult to show that if P is localizable then $\mathcal{C}(P) \subseteq \mathcal{C}(0)$. (If you don't like this step, don't worry. We shall only specialize when R is a domain, in which case $\mathcal{C}(P) \subseteq \mathcal{C}(0)$ comes free.) Hence $R_{\mathcal{C}(P)}$ is a partial ring of quotients of R which contains a^{-1}. We have the diagram

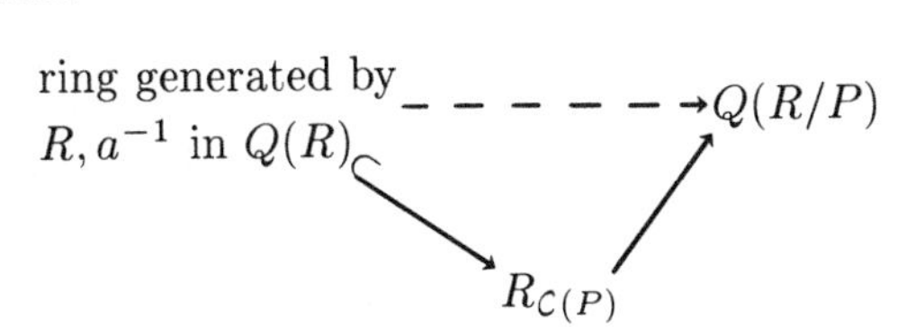

Now one either inducts on the image of the matrices in $\mathrm{GL}(t, Q(R/P))$ or finds enough primes P to specialize at so that complete information is pulled back to $A_1, \ldots, A_m$.

I'll follow Ken Brown's exposition of Lichtman's theorem.

SPECIALIZATION THEOREM (Brown). *Let k be a finite field or $\mathbb{Z}$, and let G be an infinite polycyclic-by-finite group. Then there is a normal subgroup G_0 of finite index in G such that:*

> *If I is a faithful prime ideal of $k[G_0]$ and $x_1, \ldots, x_n \in k[G_0]\backslash I$ are given, then there exists a prime ideal P properly containing I such that P/I has AR and is localizable in $k[G_0]/I$ and such that $x_i \in \mathcal{C}(P)$ for $i = 1, 2, \ldots, n$.*

Moreover, if $\operatorname{char} k[G_0]/I = 0$ *then we can find so many such P that* $\operatorname{char} k[G_0]/P$ *ranges over infinitely many rational primes.*

Proof. The idea is to control I with an abelian subgroup algebra in such a way that there is still a lot of room to maneuver over the intersection. Some initial reductions: First off, drop down by finite index so that G_0 is orbitally sound, $\Delta(G_0)$ is central and torsion free abelian, and G_0/Δ is poly-(infinite cyclic). By the Prime Controller, $I = (I \cap k[\Delta])k[G_0]$.

We divide the proof into two cases. First assume $\Delta \neq 1$. Then $I \cap k[\Delta]$ is not a maximal ideal of $k[\Delta]$. (Due to our choice of k, the ordinary Nullstellensatz implies that maximal ideals of $k[\Delta]$ are actually cofinite. No such ideals can be faithful for G_0.) Obviously $k[\Delta]$ is a Jacobson ring: $I \cap k[\Delta]$ is the intersection of the maximal ideals living above it and the corresponding residue fields run through infinitely many characteristics if $\operatorname{char} k[\Delta]/I \cap k[\Delta] = 0$. If $\mathcal{M}$ is one of these maximal ideals then $\mathcal{M}k[G_0]$, being a centrally generated ideal of $k[G_0]$, has AR. A twisted polynomial argument yields $\mathcal{M}k[G_0]$ completely prime. We wish to pick an $\mathcal{M}$ which "avoids" the x_i. Choose coset representatives g_i for G_0/Δ and write $x_i = \sum_j \lambda_j(i) g_j$ for some $\lambda_j(i) \in k[\Delta]$. Since $x_i \notin I$ we can find a $\lambda_j(i) \notin I \cap k[\Delta]$ for each i; let λ be the product of these alien elements for $i = 1, \ldots, n$. We have $\lambda \notin I$ (this from the primality of I and the centrality of $k[\Delta]$) and so we can find one of our maximal ideals $\mathcal{M}$ with $\lambda \notin \mathcal{M}$. (In the case $k = \mathbb{Z}$ we need only avoid the finitely many characteristics which divide the coefficients of λ.) For each i, we have arranged that $\lambda_j(i)$ misses $\mathcal{M}$ or, equivalently, that $x_i \notin \mathcal{M}k[G_0]$. Set $P = \mathcal{M}k[G_0]$.

Next assume $\Delta = 1$. By the Prime Controller, $I = 0$. In the case that $k = \mathbb{Z}$ we can let P be the prime ideal generated by any of the rational primes which does not divide any coefficient of $x_1, \ldots, x_n$. If k is a finite field we invoke an earlier result of Roseblade as follows. If A is a plinth for G_0 (assumed to exist by finite index finagling) and λ is as in the previous paragraph (replace Δ with A) then there is a maximal ideal $\mathcal{M}$ of $k[A]$ which contains no G_0-conjugate of λ. With a twisted polynomial argument one proves that the intersection of the finitely many G_0-conjugates of $\mathcal{M}$, when blown up to $k[G_0]$, is a desired P. With our choice of $\mathcal{M}$, none of the elements x_i lie in any conjugate of $\mathcal{M}k[G_0]$. This is enough to ensure that $x_i \in \mathcal{C}(P)$. $\square$

We turn to the Burnside problem. Since we are concerned with matrices over the ring of quotients of a prime image of $k[G]$, we will freely descend to a subgroup of finite index (e.g., when using the Specialization Theorem). The only cost will be enlarging the size of the matrices and looking at periodic subgroups of finite products.

Assume that we are given a finitely generated periodic group of matrices with entries in $Q(k[G]/I)$. A commutative specialization argument allows us to assume that k is either $\mathbb{Z}$ or $\mathbb{Z}/(p)$. We might as well assume that I is a faithful prime ideal of $k[G]$ and that G is G_0 from the Specialization Theorem. We can find a prime ideal P properly containing I and a homomorphism from the given periodic group to the group of invertible matrices over $Q(k[G]/P)$. By induction (on Krull dimension and Hirsch number), the image is finite. The group-theoretic kernel is a periodic group which lies in $1 + \operatorname{Mat}_n(\overline{P}_{\mathcal{C}(\overline{P})})$, where the overbar indicates the image in $k[G]/I$.

By construction, $\operatorname{char} k[G]/P$ is a positive prime q. We claim that every element in this periodic kernel has order a power of q. Suppose $x = 1 + \mu$, for $\mu \in \operatorname{Mat}_n(\overline{P}_{\mathcal{C}(\overline{P})})$, has order r relatively prime to q. Then $(1 + \mu)^r = 1$ implies

$r\mu$ is an integer combination of higher powers of μ. Since r lies in $\mathcal{C}(\overline{P})$, it is invertible in $\overline{k[G]}_{\mathcal{C}(\overline{P})}$. Hence μ itself is a combination of higher powers of μ. Thus

$$\mu \in \bigcap_d \{\mathrm{Mat}_n(\overline{P}_{\mathcal{C}(\overline{P})}\}^d.$$

But the Specialization Theorem produced P satisfying the AR property which, in turn, forces $\mathrm{Mat}_n(\overline{P}_{\mathcal{C}(\overline{P})})$ to have AR. Therefore μ is zero by the Krull intersection theorem

Where are we? The periodic group has a subgroup of finite index which is a q-group for $q = \mathrm{char}\, k[G]/P$. If it happens that the characteristic of $k[G]/I$ is zero then we can do the same thing all over again with a prime ideal P' having $q' = \mathrm{char}\, k[G]/P'$ distinct from q. Since an element in a group cannot have an order which is the power of two different primes, we see that the periodic group is finite.

If the characteristic of $k[G]/I$ is already q, then restrict attention to the finitely generated subgroup of finite index we have described above. Its elements are of the form $x = 1 + \mu$ where μ is a nilpotent matrix. Finiteness this time follows from another (by no means obvious) specialization argument about subrings spanned by nilpotent matrices.

Valuable Rings

Of all of the commutative techniques which have been borrowed and adapted for the analysis of Noetherian group algebras, the one which appeared the latest and which may be the most fruitful, is valuation theory. Crudely stated, the group in the group algebra can be recovered, to some extent, as a "value group".

This context inspires us to use multiplicative notation for valuations. Assume that Γ is a totally ordered, possibly noncommutative, group. By a valuation on the ring R taking its values in Γ, we mean a function $v\colon R \to \Gamma \cup \{\infty\}$ such that

$$\begin{aligned} &\text{(i)} && v(a) = \infty \Leftrightarrow a = 0, \\ &\text{(ii)} && v(ab) = v(a)v(b), \\ &\text{(iii)} && v(a+b) \geq \min\{v(a), v(b)\}. \end{aligned}$$

When R is a k-algebra, then we say that v is a k-valuation if $v(\lambda) = 1$ for all $\lambda \neq 0$ in k.

As our first application of valuations, we'll finally prove Bergman's Theorem. The idea is that if P is a prime ideal of the group algebra $k[A]$ (where A is a finitely generated abelian group) then there is a torsion free (and hence orderable) subgroup $B \subseteq A$ such that $k[B]$ imbeds in $k[A]/P$, and the field of fractions of the latter is a finite extension of the field of fractions of the former. The subgroup B is a value group for $k[A]/P$ which is "naturally" associated with the group A.

The argument ahead is a hybrid of Bergman's and Passman's. It is based on a version of the König Graph Theorem which is undoubtedly somewhere in the literature. I just don't know where to find it. (A proof appears in an appendix to this section.)

Suppose $(X, \leq)$ is a partially ordered set. As one might expect, X is said to be Noetherian if every ascending chain $x_1 \leq x_2 \leq x_3 \leq \cdots$ eventually stabilizes. An antichain in X is a set of pair-wise incomparable elements. The collection of antichains in X can be partially ordered by requiring $A \prec B$ to mean that for each $b \in B$ one can find an $a \in A$ with $a \leq b$.

LEMMA. *If a partially ordered set is Noetherian so is the partially ordered set of its finite antichains.*

Next assume that X is a semilattice: each pair of elements has a least upper bound. If A and B are finite antichains then $A \vee B$ is the set of minimal members in $\{a \vee b \mid a \in A \text{ and } b \in B\}$. Consequently the finite antichains determine a semilattice. The lemma has an immediate corollary.

THEOREM. *The collection of finite antichains in a Noetherian semilattice itself comprises a Noetherian semilattice. In particular, any list of finite antichains has a least upper bound.*

BERGMAN'S THEOREM ([1]). *Suppose that A is a plinth for the action of a group G. Then all nonzero G-invariant ideals of $k[A]$ have finite codimension over k.*

Proof. A is a finitely generated abelian group. With the relation "inclusion", the collection of subgroups of A makes up a Noetherian semilattice.

Let P be a nonzero proper invariant prime ideal of $k[A]$. (One can always reduce to this case by looking at the finitely many minimal primes over an arbitrary ideal and noting that each is invariant under a subgroup of finite index in G.) For each nonzero α in P define $\mathcal{C}(\alpha)$ to be the set of minimal subgroups among the set

$$\{\langle ab^{-1}\rangle \mid a \neq b \text{ are both in } \operatorname{supp}(\alpha)\}.$$

Then $\mathcal{C}(\alpha)$ is a finite antichain. Set $\mathcal{C} = \bigvee_{0\neq\alpha\in P} \mathcal{C}(\alpha)$, a finite antichain of subgroups of A. Obviously, the subgroups in $\mathcal{C}$ are permuted by G.

Let the overbar denote the canonical homomorphism $k[A] \to k[A]/P$ and write K for the field of fractions of $\overline{k[A]}$. We claim that if v is an abelian k-valuation on K then $\mathcal{C} \prec \{A_v\}$ where $A_v = \{a \in A \mid v(\overline{a}) = 1\}$. We must show that for each nonzero α in P, there are distinct x and y in $\operatorname{supp}(\alpha)$ such that $xy^{-1} \in A_v$. But if $\alpha = \sum_{x\in A} \lambda_x x$ then $v(\sum \lambda_x \overline{x}) = \infty$; it follows that $v(\overline{x}) = v(\overline{y})$ for some $x \neq y$.

We have yet to use the assumption that A is a plinth. But in this case, each subgroup in $\mathcal{C}$ must have finite index in A. Thus the previous paragraph implies that A_v has finite index in A for each k-valuation v. Since the valuation group is torsion free, we have $v(\overline{A}) = 1$ for all such v.

The theorem follows if we establish that $K|k$ is algebraic. If not, there must be some $a \in A$ such that $\overline{a}$ is not algebraic over k. Clearly $k(\overline{a})$ has a k-valuation which does not vanish on $\overline{a}$. By the extension theorem for (abelian) valuations, one could then construct a k-valuation w on K such that $w(\overline{a}) \neq 1$, a contradiction. □

The first use of valuations suggests two situations in which they may be exploited further. First of all, valuations seem to apply to fields of fractions as well as to the original group algebra. Second, one is tempted to look at group algebras of torsion free nilpotent groups, since these groups can be ordered.

Farkas, Schofield, Snider, and Stafford used this point of view to prove that when two finitely generated torsion free nilpotent groups have isomorphic division algebras of fractions, those groups are isomorphic [**5**]. To my knowledge, this is the first place that noncommutative valuations appear naturally in a problem. I can outline part of the solution to this problem most efficiently with the following device: suppose $v\colon R \to \Gamma \cup \{\infty\}$ is a valuation. If Λ is any group then a lift $\tilde{v}\colon R \to \Lambda \cup \{\infty\}$ is a function such that

(i) v factors through $\tilde{v}$,
(ii) $\tilde{v}(ab) = \tilde{v}(a)\tilde{v}(b)$,
(iii) $v(a) < v(b)$ implies $\tilde{v}(a+b) = \tilde{v}(a)$.

We leave it as an exercise that if R is an Ore domain with field of fractions F then v extends uniquely to a valuation w of F and $\tilde{v}$ extends to a unique lift $\tilde{w}$ of w. (From now on, we will use the same letter v for the extended valuation.)

Take a torsion free nilpotent group G and order it so that its center Z is convex. Form the partial ring of fractions $R = k[Z]^{-1}k[G]$. If we write $K = k[Z]^{-1}k[Z]$ then K is the center of R. (In fact, it's the center of the division ring of fractions of R.) Choosing a transversal T for Z in G, it is not difficult to see that each element of R can be written uniquely in the form $\sum_{g\in T} \lambda_g g$ where $\lambda_g \in K$.

There is a degree-like valuation $v\colon k[G] \to G/Z \cup \{\infty\}$ which sends a nonzero member of the ring to the image of the smallest group element in its support. (G/Z inherits the ordering on G by the convexity assumption.) As observed above, v extends to R. Define $\tilde{v}\colon R\backslash\{0\} \to K^*\cdot G$ by $v(\sum_{g\in T}\lambda_g g) = \lambda_h h$, where h is the smallest element of the transversal that actually appears in the sum. It is a matter of book-keeping to check that $\tilde{v}$ is independent of T and that $\tilde{v}$ lifts v. Extend both maps to the division ring of fractions of R, which is also the ring of fractions of $k[G]$.

We shall argue that if G_1 and G_2 are two torsion free nilpotent groups and if ϕ: Fractions $(k[G_1]) \to$ Fractions $(k[G_2])$ is an isomorphism then $K_1^* \cdot G_1$ is isomorphic to $K_2^* \cdot G_2$. Consider the composition

$$\Psi\colon K_1^* \cdot G_1 \hookrightarrow \text{Fractions}(k[G_1]) \xrightarrow{\phi} \text{Fractions}(k[G_2]) \xrightarrow{\tilde{v}_2} K_2^* \cdot G_2.$$

On the far left we have a nilpotent group. (The center of $K_1^* \cdot G_1$ is K_1^* and $K_1^* \cdot G_1/K_1^* \simeq G_1/Z_1$.) Thus to prove that Ψ is one-to-one we need only check that it is faithful on its center K_1^*. But ϕ faithfully carries the center of one division ring to the center of the other.

To see that Ψ is onto, first notice that if g and h are in T_1 and if λ_g and λ_h are in K_1^* then $(\lambda_g g)(\lambda_h h)^{-1} \notin K_1^*$. Since Ψ is one-to-one, the image of this element under Ψ cannot be in K_2^*. In other words, $v_2(\phi(\lambda_g g)) \neq v_2(\phi(\lambda_h h))$. Therefore, if $a = \sum_{g \in T_1} \lambda_g g$ is a nonzero member of $k[G_1]$ with each coefficient $\lambda_g \in K_1^*$, then condition (iii) in the definition of the lift of a valuation implies that $\tilde{v}_2(\phi(a)) = \Psi(\lambda_h h)$ for one of the summands $\lambda_h h$ of a. It follows that

$$\tilde{v}_2 \circ \phi(\text{Fractions}(k[G_1])^*) \subseteq \Psi(K_1^* \cdot G_1).$$

Rephrasing this, $K_2^* \cdot G_2 \subseteq \operatorname{Im} \Psi$. We have shown that Ψ is an isomorphism.

As an immediate corollary

$$G_1/Z_1 \simeq (K_1^* \cdot G_1)/K_1^* \simeq (K_2^* \cdot G_2)/K_2^* \simeq G_2/Z_2.$$

This is not far from the more desired conclusion that G_1 and G_2 are isomorphic. In the case of finitely generated groups it is possible to use a cancellation argument together with a second application of valuations to get there.

Appendix II

We begin with a partially ordered set $(X, \leq)$. Define $F(X)$ to be the collection of finite nonempty antichains in X. For A and B in $F(X)$ we have written $A \prec B$ to mean that for each $b \in B$ there exists an $a \in A$ such that $a \leq b$. This relation is a partial order. (Only the verification of antisymmetry takes advantage of the fact that we are comparing antichains.)

PROPOSITION. *If $(X, \leq)$ is a Noetherian poset then $(F(X), \prec)$ is a Noetherian poset.*

Proof. Suppose that $A_1 \prec A_2 \prec A_3 \prec \cdots$ is an ascending chain in $F(X)$. We construct a directed graph out of the union of points in A_i as follows. For points x and y draw $\overset{x}{\underset{y}{\uparrow}}$ if there exists an $n > 0$ with $x \in A_{n+1}$, $y \in A_n$, and $y < x$. Call this graph $\mathcal{G}$. It is clear from the strictly increasing nature of the chain that $\mathcal{G}$ is infinite.

Step I. If m is a vertex of $\mathcal{G}$ which lies in infinitely many A_i then m is maximal in $\mathcal{G}$.

Let N be the smallest positive integer such that $m \in A_N$. Suppose that m also lies in A_j for some $j > N$. Since $A_N \prec A_{N+1} \prec \cdots \prec A_j$ we can find $x_d \in A_d$ with $m = x_j \geq x_{j-1} \geq \cdots \geq x_N$. Thus m and x_N are comparable elements of A_N. We conclude that $x_N = m$ and hence that m lies in every A_t for $t \geq N$. In particular, if z lies above m in $\mathcal{G}$ then z and m will be comparable elements of one such A_t.

Step II. Every vertex in $\mathcal{G}$ has finite out-degree.

By the first step, we need only show that if x lies in the *finite* list $A_{f(1)}, \ldots,$ $A_{f(n)}$ (and in no other A_i) then its out-degree is finite. But the target of any arrow from x ends at one of the points in $A_{f(1)+1}, \ldots,$ or $A_{f(n)+1}$. There are only finitely many available end-points.

Step III. $\mathcal{G}$ has arbitrarily long finite chains.

Define the "level" of a vertex inductively: members of A_1 are in level one. A vertex lies in level $n+1$ if it covers a vertex in level n. By Step II, each level has finitely many vertices. Now any chain of length M in $\mathcal{G}$ which begins at the first level has vertices only in the first M levels. Consequently, if $\mathcal{G}$ has a bound on its chain length then $\mathcal{G}$ is finite—a contradiction.

Step IV. $\mathcal{G}$, and hence X, has an infinite ascending chain.

Given the previous two steps, this is just a statement of the König Graph Theorem. □

Generally speaking, $F(X)$ is better behaved than X. For instance, when $(X, \leq)$ is a semilattice, Dilworth [**3**] essentially observed that $(F(X), \prec)$ is a distributive lattice. Here $A \wedge B$ is the collection of minimal elements in the set-theoretic union of A and B.

REFERENCES

[1] Bergman, G. M., *The logarithmic limit-set of an algebraic variety*, Trans. Amer. Math. Soc. **157** (1971), 459–469.

[2] Brown, K. A., Lenagan, T. H., and Stafford, J. T., *K-theory and stable structure of some Noetherian group rings*, Proc. London Math. Soc. (3) **42** (1981), 193–230.

[3] Dilworth, R. P., *Some combinatorial problems on partially ordered sets*, Proc. Tenth Symp. in Appl. Math., Amer. Math. Soc., Providence (1960), 85–90.

[4] Donkin, S., *Locally finite representations of polycyclic-by-finite groups*, Proc. London Math. Soc. (3) **44** (1982), 333–349.

[5] Farkas, D. R., Schofield, A. H., Snider, R. L., and Stafford, J. T., *The isomorphism question for division rings of group rings*, Proc. Amer. Math. Soc. **85** (1982), 327–330.

[6] Hall, P., *Finiteness conditions for soluble groups*, Proc. London Math. Soc. (3) **4** (1954), 419–436.

[7] Hall, P., *On the finiteness of certain soluble groups*, Proc. London Math. Soc. (3) **9** (1959), 595–622.

[8] Harper, D. L., *Primitivity in representations of polycyclic groups*, Math. Proc. Camb. Phil. Soc. **88** (1980), 15–31.

[9] Jategaonkar, A. V., *Certain injectives are Artinian*, Noncommutative Ring Theory, Kent State 1975, Springer Lecture Notes in Math. **545** (1976), 128–139.

[10] Jategaonkar, A. V., *Integral group rings of polycyclic-by-finite groups*, J. Pure Appl. Algebra **4** (1974), 337–343.

[11] Jategaonkar, A. V., *Localization in Noetherian rings*, London Math. Soc. Lecture Notes 98, Cambridge Univ. Press, 1985.

[12] Jategaonkar, A. V., *Morita duality and Noetherian rings*, J. Algebra **69** (1981), 358–371.

[13] Lichtman, A. I., *On linear groups over a field of fractions of a polycyclic group ring*, Israel J. Math. **42** (1982), 318–326.

[14] Musson, I. M., *Injective modules for group rings of polycyclic groups I*, Quart. J. Math. Oxford **31** (1980), 429–448 (Part II, ibid., 449–466).

[15] Musson, I. M., *Irreducible modules for polycyclic group algebras*, Can. J. Math. **33** (1981), 901–914.

[16] Passman, D. S., *The Algebraic Structure of Group Rings*, Wiley-Interscience, New York, 1977.

[17] Passman, D. S., *Universal fields of fractions for polycyclic group algebras*, Glasgow Math. J. **23** (1982), 103–113.

[18] Robinson, D. J. S., *A theorem on finitely generated hyperabelian groups*, Invent. Math. **10** (1970), 38–43.

[19] Roseblade, J. E., *Group rings of polycyclic groups*, J. Pure Appl. Alg. **3** (1973), 307–328.

[20] Roseblade, J. E., *Applications of the Artin-Rees lemma to group rings*, Symp. Math. **17** (1976), 471–478.

[21] Roseblade, J. E., *Prime ideals in group rings of polycyclic groups*, Proc. London Math. Soc. (3) **36** (1978), 385–447 (Corr., ibid **38** (1979), 216–218).

[22] Roseblade, J. E. and Smith, P. F., *A note on hypercentral group rings*, J. London Math. Soc. (2) **13** (1976), 183–190.

[23] Segal, D., *On the residual simplicity of certain modules*, Proc. London Math. Soc. (3) **34** (1977), 327–353.

[24] Smith, P. F., *The AR property and chain conditions in group rings*, Israel Math. J. **32** (1979), 131–144.

[25] Snider, R. L., *The zero divisor conjecture for some solvable groups*, Pac. J. Math. **90** (1980), 191–196.

[26] Wehrfritz, B. A. F., *Groups of automorphisms of soluble groups*, Proc. London Math. Soc. (3) **20** (1970), 101–122.

[27] Zalesskii, A. E., *On the semisimplicity of a modular group algebra of a solvable group*, Soviet Math. **14** (1973), 101–105.